STP 1422

Performance of Exterior Building Walls

Paul G. Johnson, editor

ASTM Stock Number: STP1422

ASTM International
100 Barr Harbor Drive
PO Box C700
West Conshohocken, PA 19428-2959

Printed in the U.S.A.

Library of Congress Cataloging-in-Publication Data

Symposium on Performance of Exterior Building Walls (2001 : Phoenix, Ariz.)
Performance of exterior building walls / Paul G. Johnson, editor.
p. cm. — (STP ; 1422)
"The Symposium on Performance of Exterior Building Walls was held in Phoenix, Arizona on 31 March–1 April 2001"—Frwd.
"ASTM stock number: STP1422."
Includes bibliographical references and index.
ISBN 0-8031-3457-6
1. Exterior walls—Congresses. I. Johnson, Paul G., 1949– II. Title. III. ASTM special technical publication ; 1422.

TH2235.S96 2001
690′.12—dc21

2003044447

Peer Review Policy

Each paper published in this volume was evaluated by two peer reviewers and at least one editor. The authors addressed all of the reviewers' comments to the satisfaction of both the technical editor(s) and the ASTM International Committee on Publications.

To make technical information available as quickly as possible, the peer-reviewed papers in this publication were prepared "camera-ready" as submitted by the authors.

The quality of the papers in this publication reflects not only the obvious efforts of the authors and the technical editor(s), but also the work of the peer reviewers. In keeping with long-standing publication practices, ASTM International maintains the anonymity of the peer reviewers. The ASTM International Committee on Publications acknowledges with appreciation their dedication and contribution of time and effort on behalf of ASTM International.

Printed in Bridgeport, NJ
2003

Foreword

The *Symposium on Performance of Exterior Building Walls* was held in Phoenix, Arizona on 31 March–1 April 2001. ASTM International Committee E06 on Performance of Buildings served as the sponsor. The symposium chairman and editor of this publication was Paul G. Johnson, Smith Group, Inc., Detroit, Michigan.

Contents

Overview

This publication is the most recent in a series resulting from symposia presented by subcommittee E06.55 between 1990 and 2001. This Symposium, "Performance of Exterior Building Walls," was held March 31 and April 1, 2001 in Phoenix, Arizona.

In each of these previous symposia a specific subject relating to exterior building walls has predominated. This symposium was different in that the call for papers invited presentations from a broader spectrum of exterior building wall issues. The primary topic was to be the *performance of exterior building walls*. Not leaks, not wind resistance, and not structural evaluation, but *performance*. One of the goals for this symposium was to show the broad spectrum of topics related to exterior building wall performance, and similarly the types of people required to accomplish the goal of good performance. This was the stated goal, to address various performance aspects of exterior building walls. The presenters did a good job of addressing various issues and a good mix of individuals representing the types of parties involved in the design and construction process participated in this symposium. Presentations were made on product development, code issues, seismic considerations, wind evaluation, methods to predict condensation, and more. The presenters included chemists, contractors, structural engineers, architects, educators, and forensic investigators among others. There were also two non-technical presentations. One was from an owner addressing the importance of effective communication. The second was from an attorney, explaining why a leak (physical) may not really be a leak (legal).

All of the presentations and the papers in this publication address ways to improve the performance of exterior building walls, or ways to identify, understand, and avoid the factors leading to failures. As can be seen in these papers, exterior building walls are subject to failure for many reasons, including errors in analysis, design, specification, fabrication, and construction. To a high degree, these failures are preventable if procedures and methods already known are followed. The information provided by this symposium and this resultant publication provides much grist for the mill of building design and construction. There is, however, a separate issue that is perhaps equal in importance to the information provided by the individual papers. There is a vast amount of solid information regarding these issues already available, and more is available every day. Why is this existing information often not applied and used? Why do so many failures continue to occur in exterior building walls, and what can be done to correct this situation? Of course this symposium did not provide all of the answers. What it did was bring together a group of individuals and provide an opportunity to present new ideas, consider old questions in different ways, and provide food for thought on how to attain better performance from exterior building walls. This is perhaps the greater value of these symposia and of these publications; the forum for discussion and a method to make the information widely available.

The members of E06.55 hope to continue with these symposia as a forum for discussion, and the STP publications as a method to record and distribute the wealth of information available to us.

Paul G. Johnson
Smith Group, Inc.
Detroit, MI

SECTION I

William J. Pierce, CPE [1]

Meeting of Minds—Architect, Contractor, and Owner, The Subtle Process of Communication. [2]

Reference: Pierce, W. J., "**Meeting of Minds — Architect, Contractor, and Owner, The Subtle Process of Communication,**" *Performance of Exterior Building Walls*, ASTM STP 1422, P.G. Johnson, Ed., ASTM International, West Conshohocken, PA, 2003.

Abstract: The façade of any structure represents collaborative efforts by architect, owner and contractor. However, these efforts sometimes result in a less than successful project. A lack of understanding around process, individual roles and project expectations appears to be the culprit. The real question, how to change the outcome for greater success. I believe that one critically important ingredient is open and honest communication between owners, architects and contractors pertaining to project expectations, scope and final results. The architect is a pivotal partner, a first stringer with understanding of design, construction methods and processes. The architect is critical to the success of the overall project. How responsive should architects be to the owner? As a partner they should educate the owner as to best methods of project delivery, construction methods and contractors suitable to deliver a mutually satisfactory project. What role does the owner expect of the architect? Is it strictly design, project management, consulting, partnership, stakeholder, educator, employee or some combination? The owner's expectations of the architect vary by project, relationship and owners real understanding of the project. The owner and architect both require clear communications in expressing the needs and true expectations of the project. Once the owner and architect understand one another, they create a process incorporating project definition, scope, project specification and selection process toward soliciting a contractor to round out and expand the owner architect partnership. This newly formed relationship of owner, architect and contractor moves forward in a collaborative manner in which each individual contribution and success complements the overall project success. Let's examine the relationship between owner, architect and contractor relative to Exterior Building Walls in obtaining maximum efficiencies, durability and longevity by improving communications from beginning to end of the project.

Keywords: Owner, architect, contractor, communications

Failures of exterior wall systems directly affect building usage and service life. These systems deserve special consideration from building owners. The following opinions apply to all aspects of the design and construction of buildings - especially the exterior building envelope, and particularly, walls. We have the knowledge, the materials, and the construction ability to avoid exterior wall system failure. So why don't we?

As is true in so many other situations, the failure of wall performance is, in my opinion, largely due to the failure to communicate effectively and properly. I believe that the number of exterior wall failures could be significantly reduced if we, the

Architect/Owner/Contractor team, working in cooperation, could solve this single problem.

The construction, reconstruction or renovation of any building, or portion thereof represents collaborative efforts by architect, owner and contractor. However, these efforts sometimes result in a less than successful project. A lack of understanding around process, individual roles and project expectations is a probable culprit.

During a lecture given by Michael Haggans, AIA, on Project Programming in Reno Nevada in 1999, a slide of a quote by architect named Willie Pena [3] was shown. Willie Pena [4] suggests, "Good Buildings don't just happen…" If this is true, how do we make it happen? This thought led me to begin a search for possible keys to consistent project success.

I began by examining the projects with which I had been involved. My project experience ranged from small renovations/retrofits to more complex construction involving both architectural and mechanical components. I reviewed them from beginning to end. In general, from design to completion, the fundamental process appears similar.

As an operational engineer, I am expected to fully understand my function. I am expected to perform in a specified manner and to expect the same of others. In progressing from operational engineering to managing operations, I am constantly forced to think differently. Now, I am responsible for designating other peoples' function and defining the parameters within which that function is to be performed. Now, it is my responsibility to be always certain that the "other guys", be they my employees or contracted professionals, are doing their jobs and doing them to my stated parameters.

This transition, from operational engineer to director, was dependant on communication. First, I had to discover the importance of communication. Then, I had to, by trial and error, become an effective communicator. Effective communications are honest and open in clarifying duties and responsibilities. This will lead to trust.

My successful projects all had excellent relationships built on trust. This trust was dependant on open and honest communication. I had found myself expecting architects, engineers, contractors and contractual personnel to understand and be able to effectively translate my needs and desires into a successful project. However, the communication skills and trust levels acquired over time were not consistent from project to project, team to team. It now became necessary to develop a level of consistency that would apply in all situations and work equally well with all disciplines.

The Beginning

Open and honest project communication must begin with the owner. The owner must have a clear vision of the project as well as the ability to share this vision with the architect. The owner must have a clear definition of project scope and desired outcome. The owner must share their expectations for the project process, its' communications, performance and outcome. The owner must be willing to understand and adjust to the fact that they may not fully understand project process. The owner must be willing to learn from others and allow the project to evolve.

First Step

The owners' open and honest communication starts with architect selection. The owner must clearly define the performance expectations and roles of the architect which may include many levels of service such as:

- Strictly a design service.
- Project manager, overseeing the project for the owner.
- Consultant, checking the validity of proposed designs.
- Partner, stakeholder where the A/E firm has a vested interest in the projects' success.
- Educator, assisting the owner in making decisions regarding the process and ultimate product.
- Employee, acting solely at the command of the owner.

Obviously, the architect's bid and any subsequent contract will confirm his understanding and acceptance of these expectations. Clear definitions of project budget, schedule and resource availability are the reality check. Owner and architect must be in agreement. Is the project properly budgeted? Is the schedule feasible? Are resources available? Appropriately answering these questions is the first test of the owners' and architect's open and honest communications.

Once the architect's role has been defined, candid discussions about the financial relationship including fees and project budget are imperative. A good contractual relationship to clarify design fees, percentage of project budget, construction management is a critical element in the successful project. The American Institute of Architects has standard documents available that can be utilized as a foundation for defining these contractual relationships including design service, project management and consulting.

Refining the Owners Vision

Winston Churchill addressed the critical nature of structural aesthetics in his comment, "We shape our buildings; thereafter they shape us" [5]. The owners' vision should be a building that enhances his image yet is clearly recognizable within the community.

The architect must begin to refine this vision into a workable project. Working with the owner, the architect must take the raw vision through a series of efforts that educate the owner with regard to his expectations. Discussions about the aesthetics of the project must take priority. Quality, maintainability, initial cost and cost of ownership are among the issues to be resolved. Explanations of the merits of various systems should also be provided.

The owners' requirements include his expectations.

- Quality & Performance - Maximum performance and durability based on the criteria available for exterior wall weatherproofing.
- Function & Longevity - Ability to extend beyond the normal life cycle.
- Maintainability - The structures' ability to be weathertight and good looking throughout the structures' life cycle at reasonable cost.
- Aesthetics - Appearance reflects the owners' intent.
- Schedule and budget – The project meets timing and financial requirements.

The owner's faith in the architect's ability is essential for a successful project. However, the owner must be willing to educate himself as to basic wall construction and to challenge the conclusions of the architect. Challenging the architect is not adversarial. It is affirmation of reality.

Now, the architect begins indoctrinating the owner in the process of design. The commitment to open and honest communication is tested as this process unfolds. The thousand and one questions regarding plan reviews, impacts of code, finishes, lighting and equipment selection serve as reminders of the need for superior communication. The architect becomes an educator and mentor during this process, serving as guide and advisor to the owner. The owner must acknowledge that the architect has the lead role during this phase of the project.

Project Delivery

The architect, understanding design and construction methods, advises the owner as to the best avenues of project delivery. Owner and architect must agree on the best delivery method that meets all the project goals and objectives. Project delivery can be one of several methods:

- Owner acting as a General Contractor.
- Construction Management.
- General Contractor.
- Project Management.
- Design/Build.
- Fast Track Design/Build.

Contractor Selection

Contractor selection evolves as the next project step. Communication with contractors can be clouded by the misconception that contractors generally are only interested in maximized profit and minimized product. How do we, as partners (owner and architect), accept a new partner or partners in the process? We must reduce our preconceived adversarial notions of contractors for a successful project. Assimilating the contractor into the existing communications process can blunt the fundamental adversarial nature of architect/owner versus contractor.

One possible problem to a clear communication process can be the project's contractual relationship. The owner and the architect have a separate contract and the general contractor and owner have their own contractual agreement. The owner is responsible for these two separate contracts. It is imperative that the owner review these contracts for areas of overlap or possible conflict. The coordination of contracts is essential as one or the other may inadvertently create a problem in the relationship. Then too, in the evolution of the project, unforeseen conditions, work and scope may not be covered by the basic contractual relationship of the parties. Therefore, a method of conflict resolution must be established.

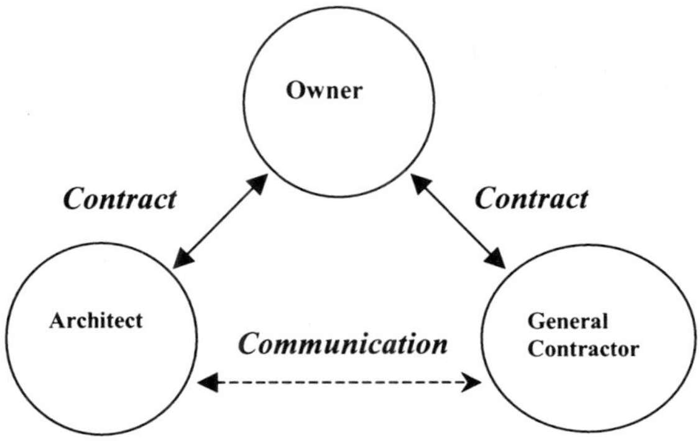

Project Communication Diagram

Again, the contractor's role within the clearly defined scope and outcome of the project must be stated at the outset both for bid preparation and again in the final contract. The contractor's role and responsibilities should be clearly identified in the construction contract documents. All details regarding the financial aspects of the relationship must be addressed in each contract. Then, a communication hierarchy must be established to accommodate and facilitate the roles of the owner, architect and contractor during the project. This hierarchy represents the formal contractual issues and informal daily communication necessary for mutual success.

A clear set of drawings and specifications is required. Not the standard boilerplate but a composite of the owners' requirements and the architect's experience should be embodied in these documents. Coordination of drawings and specifications is critical. However, this may be an area of contractual conflict for the parties. It may be useful to all to have a neutral party review drawings and specifications to keep open and honest communications flowing. An architectural professional not associated with the project may perform this independent review. The owner employs this professional. Prior to any independent review all must agree or understand it is part of the process.

Construction

Now, the roles of the owner, architect and contractor have been clearly defined, understood and agreed upon by all parties. The parameters of each party's function have been clearly outlined.

As construction begins, myriad questions concerning specifications, materials, schedule, coordination drawings, site preparation and other legitimate concerns test the commitment to communication. Good communications are based on trust that all are proceeding with the projects' successful outcome in mind. Standard weekly meetings will assure continuity. However, specific or focused meetings will resolve major problems, especially as they arise. Fundamental problem resolution searches for workable solutions without laying blame at someone's feet. Resolving issues quickly reaffirms commitment to the project and its partners. This is where walking the talk is critical. Timeliness is imperative.

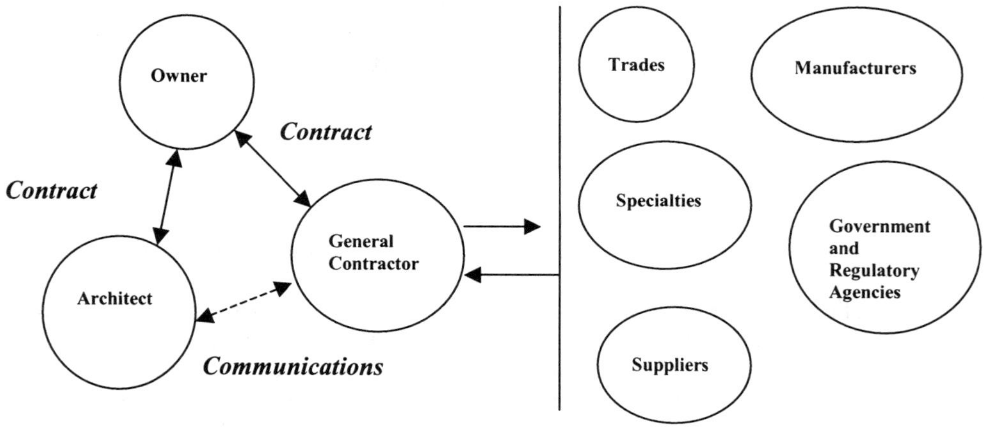

General Contractor Communication Diagram

Communications between the general contractor, his suppliers, trades and manufacturers have direct and indirect impact on the project. The general contractor should provide the input of these additional players relative to schedule, budget and the occasional technical issues. This resource creates opportunities for possible alternative products and methods while providing unique problem solving abilities. Honest communication is clear about expectations, open to alternative solutions and committed to a successful project as well as participant's mutual success.

Finally

In real estate it's called curb appeal. The façade or exterior should communicate, at least in part, the nature - structurally and professionally – of that building. This requires the dedicated cooperation of the owner, architect and contractor responsible for integrating the appropriate walls. Cooperation of that magnitude can only be facilitated by communication.

There is nothing new or revolutionary in recognizing the importance of communication. Unfortunately, it's easy to overlook the obvious. Communications and trust can result in a better project. Too often the owners, architects, engineers and contractors revert to the ingrained belief that they should do their jobs and let others do theirs. This is possible only if someone has spelled out those jobs to everyone involved.

Owners, architects, engineers and contractors all too often retreat to the learned responses of a contractual situation. In this instance that would translate as: the contractor is the problem; the owner is the problem; the A/E firm is the problem. These tendencies do not serve the project. Allowing the everyday "stuff" of a project to overwhelm the greater picture must be avoided. Only through the diligent pursuit of a relationship based on open and honest communication between the owner, architect and contractor can a successful project be achieved.

References

[1] Pierce, W. J. ,Certified Plant Engineer, Director, Building Operations, The Detroit Institute of Arts, Detroit, MI, 48202.
[2] This article is a non-technical perspective of communications related to the construction process.
[3] Pena, W., "Problem Seeking," an Architectural Program Primer, 1969.
[4] Quote from APPA seminar on project programming lecture by Michael Haggans, AIA, Reno NV, 1999.
[5] Quote from APPA seminar on project programming lecture by Michael Haggans, AIA, Reno NV, 1999.

Robert J. Kudder,[1] *Kenneth M. Lies,*[1] *and Brian A. Faith*[2]

Ambiguities, Changes, and Contradictions in Building Wall Literature

Reference: Kudder, R. J., Lies, K. M., and Faith, B. A., **"Ambiguities, Changes, and Contradictions in Building Wall Literature,"** *Performance of Exterior Building Walls, ASTM STP 1422*, P. G. Johnson, Ed., ASTM International, West Conshohocken, PA, 2003.

Abstract: There is an enormous body of information about the behavior of building walls and numerous guidelines, codes and standards to assist designers in establishing wall performance criteria, selecting and specifying wall materials, and testing to verify wall performance. There is so much information available that it is difficult for a designer to be familiar with and to digest all of it. Guidelines also change over time, often in a way which significantly changes the meaning of performance criteria. In addition, the nomenclature used in this body of information is not clearly and consistently defined. This can make a designer's task difficult, necessitating attention to the current meaning of the guidelines and how changes could impact a design. Examples of ambiguities and contradictions in standards and industry practices are discussed.

Keywords: building walls, wall types, water infiltration resistance, leakage

The building envelope design process is guided by an enormous body of standards, codes, technical publications and product information. For example, design load criteria are given in ASCE Minimum Design Loads for Buildings and Other Structures (ASCE 7-88) as well as the model and local building codes. Product and component performance criteria are given in the national standards published by industry organizations such as The American Architectural Manufacturers Association, The National Roofing Contractors Association and The Brick Industry Association. General application guidance and specific recommendations for detailing and assembly are given in manufacturer's product literature, along with performance expectations for specific products. Test procedures for evaluating performance and for quality assurance are published by consensus organizations such as The American Society for Testing and Materials (ASTM) and The American National Standards Institute.

[1] Principal and [2] Associate, Raths, Raths & Johnson, Inc., 835 Midway Drive, Willowbrook, IL 60521.

ASTM also publishes Standard Guides and Standard Practices which include recommendations for design and construction practices. Government and research organizations such as The National Institute for Science and Technology, Oak Ridge National Laboratory and the National Research Council of Canada disseminate information on new technologies and research results related to wall performance and durability. Current information and case studies are disseminated by professional publications such as *The Specifier*, *The APT Bulletin*, *The Masonry Society Journal* and ASTM symposium proceedings (STPs) and Manuals (MNLs). For the consumer and contractor audience, information is disseminated by publications such as *Fine Homebuilding* and *The Journal of Light Construction.* In addition, textbooks on the design and behavior of the building envelope and web sites dealing with wall materials, products and construction are now readily available.

There is such an abundance of information about the building envelope that a designer must selectively seek out and digest information applicable to a particular project. It is difficult to imagine a designer being familiar and digesting all of the information available for all of the various components and systems in the building envelope. In addition to its shear volume, the body of design information is constantly evolving and nomenclature used is often unclear. The authors have been surprised and disappointed by the ambiguities and contradictions encountered while trying to understand the meaning and intent of current design guidelines. This paper presents several concepts encountered by the authors which were found to be confusing and which may interfere with the optimal design of a building envelope.

Classifying a Wall Type

Generically identifying a wall type is a seemingly simple task, but is actually extremely difficult. At the 1995 annual meeting of The Masonry Society, Rochelle Jaffe [3] conducted a survey in which the attendees were asked to describe a series of walls represented by cross-sectional drawings. The data from this "name the wall" exercise were tabulated and reported at the meeting, and the results were illuminating. Almost every imaginable permutation of descriptor terms such as "cavity," "drainage," "barrier," "veneer," and "composite," etc. were used by the attendees to describe each of the walls. Clearly there was no real consensus about the best descriptor for the example walls, even among the specialists attending the meeting. After much discussion, it became apparent that there was a general consensus about the definition of each of the descriptor terms when addressed in an abstract, isolated manner. Divergence occurred in applying the terms in the context of a particular wall. Apparently, each participant in the survey focused on a particular aspect of the wall, and used that aspect to characterize the overall behavior of the wall. Since a wall can have many different components and a combination of behavior characteristics, it is understandable that different specialists with different interests and experiences might identify the walls in different ways.

One response to the ambiguity in classifying a wall has been proposed by Clayford

---

[3] Principal Engineer, Construction Technologies Laboratory, Skokie, IL.

Grimm,[4] and the concept has been recommended for evaluated by the ASTM E06.55 subcommittee. He proposes that a series of standardized wall designs for a variety of straightforward applications be developed by some consensus organization and that each design be given an alphanumeric identifier. There is a precedence for this approach. Underwriter's Laboratory (UL) publishes a manual of fire-rated assemblies, each of which is given a simple alphanumeric code for identification. UL apparently does not perceive a need to apply descriptive labels or names. The Tile Council of America uses a similar coded classification system for various floor and wall installations, forgoing the use of narrative descriptors. If the reader has ever participated in an ASTM Taskgroup meeting while definitions were being debated, this approach might seem very attractive.

The authors believe that the opposite approach to classifying walls would be more useful. Rather than substituting a single alphanumeric code for a wall descriptor name, the number of descriptors should be increased. Perhaps the difficulty in arriving at a consensus classification or description for a wall results from an presumption that one descriptor can do the job. For example, a wall might rely on one mechanism for resisting water infiltration at its outermost surface and a different mechanism for resisting migration of water once it is within the wall. Furthermore, a wall is typically a combination of various components, and each component may have a different intended mechanism for resisting water infiltration. For example, the field of a wall may function reliably as a surface-sealed barrier system. The windows within the wall usually will not function reliably as a surface-sealed barrier and will require a flashing system. The interface between the field and the windows may require a double seal to function reliably. How can such a wall be described? Is it a barrier wall based on the characteristics of the field or is it a drainage system with a secondary water resistance mechanism based on the characteristics of the flashed windows? The difficulty in selecting a single descriptor for this wall is clear.

A more fundamental question is whether a single descriptor is actually necessary or useful. The authors believe that an accurate way to describe the wall is necessary and that an effort must be made to reach consensus on what the descriptors should be. After all, how we describe something reflects and affects how we think about it in the design process. If this entire wall were described solely as a surface-sealed barrier, the description would indicate a flawed understanding of the behavior of the overall system. Accepting the surface-sealed barrier descriptor for the overall system could lead to deficient detailing of the window. Accepting the redundant or secondary water resistance mechanism descriptor for the overall system could lead to excessive and unnecessary redundancy for the field of the wall where it is not necessary. For this wall, more than one descriptor is needed, such as "surface-sealed barrier with drainage at discrete water infiltration sources (fenestration, penetrations, etc.) and double seals at the interfaces." Even though it is lengthy, this descriptor actually describes how the wall functions to resist water infiltration, and does not contribute to misunderstanding about behavior or inappropriate design decisions.

---

[4] Consulting Architectural Engineer, Austin, TX.

Flashing and "Redundancy"

The concepts of redundancy and primary/secondary water infiltration resistant barriers are critically important to good wall performance, yet these terms are currently used in such a wide variety of ways that their meaning has become ambiguous. Debates have raged about whether primary means the most important barrier, or the first barrier encountered, or the barrier which does the most to resist infiltration, or the barrier without which the wall cannot function. For example, in a wall clad with conventional stucco applied over building paper on non-moisture-resistant wood sheathing and studs as recommended by industry standards, is the stucco or the paper the "primary" barrier? If the meaning of "primary" is the most important component, without which the system is not viable, then the building paper is the "primary" barrier. This would imply that the stucco is the "secondary" barrier whose purpose is to protect the paper. There is actually no expectation that the system will work without the building paper, so arguments about which is "primary" and which is "secondary" is moot - both the stucco and the building paper are required, and together they constitute the "barrier". Unfortunately, attempting to apply the terms "primary" and "secondary" to a wall section like this can result in flawed thinking about the system because some measure of redundancy is implied, and the building paper might be thought of as available for some other purpose, such as a drainage plane.

In describing water infiltration control strategies, the meaning of "redundancy" has created confusion. The authors think of redundancy as a water infiltration resistant mechanism available as a backup in the event that the intended mechanism fails or changes over time. If a component is absolutely essential for the performance of a system, does it provide redundancy? In the stucco example above, the authors would not consider the building paper redundant.

Flashing is another wall component often thought of as providing redundancy. In the authors opinion, for flashing to provide redundancy there should be no expectation of its getting wet in normal service unless some other component fails. By this reasoning, flashing in a masonry cavity wall is not redundant. Rather, it is essential and fundamental to the acceptable behavior of the wall. In other situations, flashing can be redundant. Arguments have been made that fenestration can be considered to pass an ASTM E 331 Standard Test Method for Water Penetration of Exterior Windows, Curtain Walls and Doors by Uniform Static Air Pressure Difference test if leakage is controlled by flashing. But, who provides the flashing? If the flashing is designed, detailed and installed separately from the fenestration system, then its purpose is to provide redundancy and it should not get wet unless the fenestration fails. The fenestration should pass the infiltration test without the independently installed flashing. If flashing is essential for the performance of the fenestration, then it is not redundant and should be designed, tested and supplied as part of the fenestration system. The Owner can then evaluate the merits of providing an additional flashing or some other mechanism if redundancy is desired.

Testing and Certification

Water infiltration testing described in ASTM E 331 and its derivatives have been debated, refined and modified in the ASTM consensus process for years. Yet, it continues to be a source of confusion and revisions have to a certain extent changed how a designer can interpret successfully passing the test. One of the authors was asked to witness the water infiltration testing of large sliding glass doors conducted at the manufacturer's laboratory. The door was installed with head and jamb receptors, as required by the project documents. The project documents also required that all installation accessories and hardware be included in the qualifying performance tests because, in our mind, the purpose of the test was to verify performance of the entire door system. However, the test was set up with the interface between the receptors and the door frame taped off to remove them from exposure to water during the test. The manufacturer argued that the water and air infiltration resistance tests were intended only for the basic door unit, exclusive of installation accessories, regardless of what the project documents stated. They also argued that there should be no concern about water leakage through the installation accessories because the project design included flashing. The counter-argument was that if flashing was necessary for acceptable performance of the door system then it should be supplied with the door. The "redundancy" which the independent flashing provided should not be usurped by leaking installation receptors. In this case, even with a clear statement of the scope of the test requirements in the project documents, the industry standard test procedure could have been applied in a manner which defeated the purpose of the test.

The definition of leakage in window standards has changed over time. In the 1980 and 1987 ANSI/NWMA Industry Standard for Wood Window Units (I.S. 2-80 and I.S. 2-87) standards, water leakage was defined to include any water that flowed into the "wall area." It would seem reasonable that a designer could interpret this to mean that no water could leak through any part of the window into any part of the wall. In the 1993 issue of I.S.2-93, the definition for leakage was changed by deleting the word "area" and adding the word "cavity" after the word "wall." However, there can be confusion and differences of opinion on the interpretation of what the wall cavity is, and where it begins and ends. The language in window industry's standard has again been changed. The 1997 issue of I.S.2-97 has defined what some industry experts have called the "wet zone." In this latest standard, new terms have emerged called the "water plane" and the "test plane." A graphic in this standard (Figure 1) illustrates

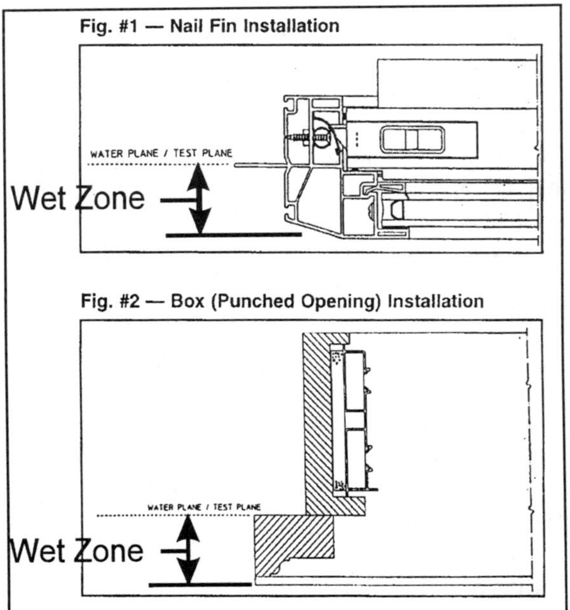

Fig. 1 - Figures #1 and # 2 from AAMA/NWWDA 101/I.S.2 - 97. Annotation by authors.

that this plane is in line with the mounting flange on windows with nailing fins and in line the backside of the brick molding on traditional wood windows. In essence, components of the windows that are exterior of this plane are not included in certification testing, or in other words, they are permitted to leak both air and water, and the window would still be considered a certified product. It is also not clear where the water plane is for windows which are installed by some method other than a nailing flange or a brick mold, such as metal straps or fasteners through the jamb and head members.

ASTM E 331 also revised the definition of water leakage. In the 1986 version, the definition for "water leakage" included water that penetrates through the frame of the test specimen. In the 1993 version of ASTM E 331, the "water leakage" definition was replaced with a "water penetration" definition which considers only the presence of water beyond the innermost projection of the test specimen and does not include the frame. Frame leakage was deleted from the definition, but water penetration through the frame is defined as a failure elsewhere in the standard unless it is contained within drained flashing, gutters and sills. The design of wall details is directly impacted by the test procedures and pass/fail criteria in the standards. It has been the authors' experience that designers and owners typically considered most windows to be watertight and historically have included flashing beneath them as a redundant feature. If the window industry standard now is that the window frame should not be considered watertight, then any flashing beneath the windows can no longer be considered redundant, but required. If a designer wants redundancy for window leakage beneath the window, can flashing which is essential for the basic performance of the window be considered redundant?

New Wall Behavior Concepts

Modern innovative wall systems with new mechanisms for resisting water infiltration have introduced a new generation of nomenclature problems. Garden [*1*] published an early paper on the rainscreen principle, in which its characteristics and potential benefits were described. Part of the definition of the rainscreen principle was "pressure equalization," which in concept results in a balance of differential pressure on the inside and outside faces of the exterior skin of the system. This concept is intuitively attractive and has been shown to work. However, using the term "pressure *equal*ization" implies equal pressures, which stated another way implies an absence of differential pressure. If the differential pressure across the exterior skin is zero, then there should be no water penetration driven past the exterior skin. Pressure equalization requires careful detailing, comparmentalization, balancing of vent areas, a rigid cavity and a tight concealed air barrier. Guidelines for designing and detailing pressure equalized systems are evolving and there is a body of data which demonstrates that equalization can actually be achieved. In the authors experience, pressure equalization is often claimed with no verification, and the total absences of water penetration past the exterior skin is often assumed. The term "pressure equalization" is not ambiguous, but unfortunately it creates expectations about wall behavior which may not be realized. This is one instance where a more general term such as "pressurization" rather than "pressure equalization" could be more useful, at least until guidelines for achieving pressure equalization are more widely understood, applied and verified.

The terms "water management," "drainage plane," "rainscreen" and "cavity" are literally in the news. Concerns about the performance of residential cladding systems have been written about in newspapers and consumer publications, discussed at conferences [*2*], been the subject of media exposes, spawned numerous web sites and been the subject of intense litigation. In watching this drama unfold, it becomes apparent that these four terms are being used interchangeably by consumers, construction professional and their attorneys. They are not synonyms, and they are not necessarily antonyms or antidotes for a "barrier." The fact that they are not necessarily mutually exclusive also contributes to the ambiguity in the use of these terms. They are very useful terms, and if used correctly can accurately describe complex wall behaviors. The only way to use them correctly is to first understand the behavior of a particular wall configuration and then apply one of the terms, rather than apply one of the terms and assume that the wall behaves accordingly.

Need for Clarity

There is obviously a need for clear, agreed-upon identifiers for various types of walls and for various wall behaviors. The identifiers need to be unambiguous in the context of actual wall behavior, based on an understanding of the operative water resistance and control mechanisms rather than a perceived behavior based on a label. It may in fact be futile to try to use single-term or hyphenated descriptors as identifiers. A focus on nomenclature and taxonomy is diverting attention and energy from the more fundamental objective of actually understanding how a wall works. The current situation is not really analogous to the Tower of Babel. Building scientists and designers seem to understand each other and to agree on the abstract definition of most wall descriptors when considered outside the context of an actual wall. Labels and descriptors do not create behavior. If misapplied, labels and descriptors can interfere with our understanding of wall behavior, distort our thinking, and complicate the design decision making process.

References

[1] Garden, G. K., "Rain Penetration and Its Control," *Canadian Building Digest*, National Research Council, Ottawa, 1963.

[2] Carll, C., "Rainwater Intrusion in Light-Frame Building Walls," *Proceedings of the Second Annual Conference on Durability and Disaster Mitigation in Wood-Frame Housing*, Forest Products Society, Madison, WI, 2000.

David O. Prevatt, Ph.D.[1]

Wind Load Design and Performance Testing of Exterior Walls: Current Standards and Future Considerations

---

Reference: Prevatt, D. O., "**Wind Load Design and Performance Testing of Exterior Walls: Current Standards and Future Considerations,**" *Performance of Exterior Building Walls, ASTM STP 1422,* P. G. Johnson, Ed., ASTM International, West Conshohocken, PA, 2003.

Abstract: Although the main structural systems of fully engineered buildings perform adequately during extreme wind events, costly losses happen to buildings once the components of the exterior walls and claddings fail. In response to these failures, new design methods have been developed that result in higher design wind loads applied to components, and prescribe additional tests on cladding to determine the structural resistance of exterior wall elements.

This paper discusses some recent changes to the wind load design provisions of the American Society of Civil Engineers (ASCE) Standard, ASCE 7-98, that apply to exterior building walls. ASCE 7-98 includes new concepts for cladding design that consider impact resistance and topographic effects on overall wind loads. Examples compare the wind design loads obtained using ASCE 7-98 with loads obtained with 7-95 and 7-88 for regular-shaped buildings. The changes may eventually influence the exterior wall design throughout the U.S. because the recently published International Building Code (IBC-2000), formed under a partnership agreement of the three existing model building codes, has adopted ASCE 7-98.

Improving the wind performance of exterior walls depends equally on improved wind design codes as well as on improved test procedures that determine the structural capacity of installed cladding systems. The current state-of-the-art in full-scale testing of building components is discussed, and a summary of current full-scale tests is presented. The author proposes that the current fragmented design process for different cladding materials and the reliance on materials-specific performance tests is too complex and needs to be streamlined in order to improve the overall performance of building envelope systems.

Keywords: building envelope, components and cladding, wind load, missile impact, curtain walls, model building codes, windows and doors, metal edge flashing, wind-borne debris, shutters, design codes, performance testing, ASCE 7

---

[1] Senior Engineer, Simpson Gumpertz & Heger Inc., Consulting Engineers, 41 Seyon Street, Building 1, Suite 500, Waltham, MA 02453.

Introduction

The building envelope is composed of several building elements (components and cladding) that work together to shield the building's structure, its contents, and inhabitants from the elements. The components (fasteners, purlins, girts, studs, roof decking, and roof trusses) receive wind loads directly from cladding and transfer the load to the main force-resisting system. Cladding includes wall shutters, curtain walls, roof coverings, exterior windows (fixed and operable) and personnel doors, and overhead doors.

Typically, the main structural systems of engineered buildings (building structures designed by professional engineers) perform better in high winds than the components and cladding of the building envelopes. However, it is difficult to find statistics from post-hurricane investigations that support this contention because the numbers of failures due to wind loads exceeding design values, versus failures due to construction or material flaws, is not known. Researchers [*1*] found in field investigations after Hurricane Andrew that the most severe damage was in residential areas, and that most of the damage was to cladding systems. Researchers also observed cladding failures in engineered buildings, but rarely found complete structural collapses in either engineered or non-engineered buildings. Structural building systems that used cladding as part of their lateral bracing systems were most likely to suffer structural collapse.

The design of cladding for exterior walls depends upon knowledge of the strength or structural resistance of the wall systems and upon estimating the design event wind loads to which cladding will be exposed. However, some exterior wall components are sometimes not designed with the same rigor as are the main structural systems of a building. Structural designers of exterior walls use guidelines from building codes, wind load design standards, and industry literature to guide their professional judgment. The codes governing wind load design are based on models of natural wind developed using wind tunnel studies and on historical weather data. The structural resistance of building components is determined by engineering calculations, historical records of accepted performance data, and, to a limited extent, on field investigations of actual performance and damage following high-wind events.

How well is the design process functioning today? Is society being well served by the system of building design as it currently exists, and will the building constructed today have the capacity to survive the next hurricane that landfalls in the U.S.? Typically, building envelope professionals follow building code requirements developed at the local, state, and federal levels to guide the design process. The design effort is only as good as the codes themselves and the knowledge of the designer of the intent and use of the codes.

This paper reviews current wind load design standards, issues of building codes, and performance testing that relate to the exterior building walls, and the author proposes actions to improve the design process and to provide better performance for the next generation of building construction. This paper does not address workmanship and installation issues relating to the performance of building walls.

Building Codes

Model building codes provide guidelines by which loading and resistance are determined. The public implicitly expects that the design process as it is practiced today leads to improved design of the building envelope. However, the process itself may not always serve the best interests of the public. In order to satisfy the public's demand for safe building construction, the design process should include, as a minimum, the following:

1) design standards incorporating the latest knowledge based on research and observations,
2) model building codes that use information from design standards and test results of all construction materials, and
3) manufacturers of cladding materials who develop safer systems and provide test data and installation guidelines that provide the reliability the public seeks.

Unfortunately, the above constraints are not always present. Many codes and standards become outdated soon after they are published, and the reliability of published test results is sometimes suspect or lacking as manufacturers struggle to balance the economic reality of a competitive market with the public's need for accurate, up-to-date information on product performance. In addition, the design professional is required to be aware of numerous local amendments to the various building codes that vary from state to state, county to county, and town to town. This is seldom a straightforward task.

Building codes establish minimum acceptable standards for building construction, concerning public health, safety and welfare, and to protect property, and every jurisdiction adopts or authors its own building code. Currently, the U.S. has three model codes: the National Building Code, the Uniform Building Code, and the Standard Building Code, published respectively by the Building Officials and Code Administrators (BOCA), the International Conference of Building Officials (ICBO), and the Southern Building Code Congress International (SBCCI). All three of these organizations have joined to form the International Code Council, which has adopted, by reference, the American Society of Civil Engineers' Minimum Design Loads for Buildings and Other Structures (ASCE-7), a document that is revised about once every five years.

Recent changes to the model building codes and to wind load design provisions have made the wind load design process more complex, and our increased knowledge of wind/structure interactions has led to higher wind design loads. In some jurisdictions, building officials now require certification and performance testing of all building cladding elements. The increase in design effort (and costs) is related to additional engineering required to design cladding systems that have limited in-service history, are more susceptible to wind damage, and have unknown failure modes in wind events. As a result, the building envelope industry is relying more on structural test results to design wind-resistant cladding systems, although the validity of some tests may not be proven.

Building Codes and the Design of Exterior Wall Components

Wind design loading is obtained from wind tunnel studies or design wind codes, such as ASCE 7, and the structural resistance of a specific wall cladding system is determined either by testing or through calculations. For each cladding system, various safety factors are included based on historical performance of the material, past practice in the industry, or "rule of thumb." This process uses a rational approach to ensure that the structural resistance always remains above the expected loads over the life of the structure.

The design of the building envelope is a collaborative process, more so than is the structural design of a building's main structural frame. Cladding consultants, architects, manufacturers, and contractors all provide input and details in ways that sometimes cloud the lines of design responsibility. These parties rely upon building codes, standards, and performance test results in the progress of a design. The sharing and exchange of design information is quite compartmentalized by type of industry or material type, resulting in problems in obtaining uniform reliability of the material used in the assembled structure. Unlike the determination of structural resistance for traditional structural systems, calculation methods for wall components are not as straightforward in the application of flexible cladding systems as these systems undergo nonlinear geometric deformations. In addition, the wind's fluctuating behavior can induce resonance effects to flexible cladding systems that are difficult to model mathematically.

Wind Loads on Buildings Envelopes

Natural wind blowing over and around buildings creates unsteady loads from wind speeds that vary in space and in time. Sometimes the wind is a gentle breeze sufficient to ventilate and, at other times, it rises to gale force or higher. Wind speed data is recorded by anemometers typically at 33 ft (10 m) above ground at airport locations (Exposure Category C), and the wind speeds are idealized into two components, the mean wind speed and its gust effect that represents the maximum excursions of wind speed about the mean.

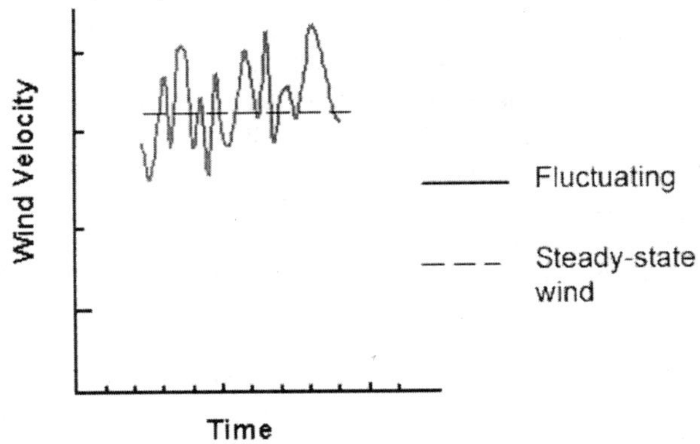

Figure 1 – *Relationship of Fluctuating and Steady-State Wind Speed*

Traditional wall cladding materials, such as brick masonry, do not respond dynamically to the fluctuating part of the wind since their natural frequencies fall outside the frequency range of the wind during extreme events. However, more flexible cladding systems, such as EIFS or flat metal panels, with natural frequencies falling within the wind frequency spectrum may experience dynamic load magnification because of their low mass and low stiffness characteristics.

The mean wind speed increases with height within the layer of atmosphere nearest to the ground up to some gradient height, Z_g. The gradient height and the slope of the velocity profile are determined by the roughness characteristics of the upstream ground surface. Rougher upstream terrain reduces the mean wind speeds and increased turbulence at all heights.

During high winds, the most vulnerable parts of the building envelope are the windows and doors. The highest wind loads are generated near corners, along eaves, and at the roof ridge for most buildings. High winds affect the exterior envelope of buildings in three ways:

1) airflow over and around a building creates external pressures and suction forces on the roof and walls;
2) wind flow into or out of the building increases and/or decreases the internal pressure; and
3) the wind blows debris (2 x 4 lumber, roof gravel, garbage cans) that may impact downstream structures, punching holes in windows or doors and breaching wall and roof claddings.

Once the building envelope is breached, higher internal pressures may combine with external pressures to significantly increase loads on the wall and roof, causing additional components to fail. Wind-driven rain entering these openings can then cause significant damage to interior building finishes and its contents.

ASCE Wind Load Design Provisions

The basic wind speed used in the ASCE 7-98 design code is the average wind speed having an annual probability of 0.02 (or fifty-year recurrence interval). In ASCE 7-95 [*2*] and 7-98 [*3*], the wind speed is determined over an averaging time of three-seconds instead of the fastest mile wind speeds used in previous versions, ASCE 7-88 [*4*] and earlier. While the two latest ASCE versions have higher basic wind speed values, this change does not increase the design wind loads; it simply reflects a different measuring system.

Wind generated pressures and forces on a structure vary as the square of the wind speed: $p = kV^2$

In addition, the frequency content of fluctuating wind speeds can magnify the pressures and forces induced on flexible structures and cladding. Wind speeds are better correlated over small areas resulting in smaller tributary areas seeing a larger pressure per unit area than larger panels. Thus, ASCE provides separate design (gust) factors for the main-force resisting systems and for the components and cladding of buildings, which, in general, have smaller tributary areas than main structural systems.

For design purposes, ASCE 7-98 defines terrain roughness (or exposure) categories as shown in Table 1:

Table 1 – *The Four Exposure Categories of ASCE 7-98*

Category	Description	Typical Location
A	Large city center with at least 50% of the building having a height in excess of 70 ft (21.3 m). Representative terrain extends upwind the greater of least 0.5 mile (0.8 km) or ten times the height of the building or structure.	Downtown of major cities, e.g., Manhattan, NY, and Chicago, IL
B	Urban and suburban areas, wooded areas, or other terrain with numerous, closely-spaced obstructions having the size of a single-family dwelling or larger. Representative terrain extends upwind the greater of 1,500 ft (457.2 m) or ten times the building height.	Downtown Greenville, SC
C	Open terrain with scattered obstructions having heights generally less than 30 ft (9.1 m). Open terrain should extend 600 ft (182.9 m) upwind.	Airports, shorelines of the Atlantic coast states, the Caribbean, Lubbock, TX, and Hawaii
D	Flat, unobstructed areas exposed to wind flowing over open water (excluding shorelines in hurricane-prone regions) for a distance of at least 1 mile (1.61 km). Extends inland from the shoreline a distance of 1,500 ft (457.2 m) or ten times the building height.	Shorelines of inland waterways, the Great Lakes, and coastal areas of California, Oregon, Washington, and Alaska

ASCE 7-98 Design Provisions

The wind load provisions of the ASCE design standard underwent major revisions with the ASCE 7-95 version, previously discussed by Smith [*5*]. Listed below are further changes to design wind loads included in ASCE 7-98 for components and cladding, listed in order from the most significant increase to the most significant decrease in wind loads.

- The topographic factor, K_{zt}, introduced in ASCE 7-95, is unchanged in ASCE 7-98. K_{zt} accounts for wind speedup at escarpments and cliffs that are isolated and unobstructed within a given terrain. The upwind distance to consider has been lengthened to the lesser of 100 times the height of the topographic feature or 2 miles (3.2 km). The topographic feature must also protrude above the height of upwind terrain features in any quadrant by a factor of two or more. (Increases load.)
- Truncated velocity pressure coefficients for low-rise buildings in Exposures B and C at the bottom 100 ft (30.5 m) and 30 ft (9.1 m), respectively. The truncation accounts for increased wind loading due to local turbulence and increased wind speeds near the surface in these rough terrains. (Increases load.)
- A wind directionality factor, K_d, is now included for buildings and other structures. This factor produces about a 5% increase in design wind loading when using factored load design, and a 15% reduction using the allowable stress design method. (Increase – Decrease.)
- The basic design wind speed map has been updated using additional information on hurricane wind speeds. The map includes predictions of hurricane wind speeds for sites away from the coasts. (Change in load varies with location.)
- The hurricane coast importance factor is not interpolated within 100 miles (160 km) of a hurricane coast. The wind speed contours are adjusted to reflect design level hurricane wind speeds. (No change in load.)

- Exposure C now includes shorelines of hurricane-prone areas and where open water extends upwind for at least 600 ft (183 m) but less than 1 mile in non-hurricane prone regions. Other open water areas remain in Exposure D. (Load decreases on hurricane shoreline.)
- Internal pressure coefficients in ASCE 7-98 are reduced to account for the imperfect correlation between the maxima of external and internal pressures. (Decreases load relative to ASCE 7-95.)

Table 2 – *Comparison of Changes in Wind Design Loads in ASCE 7*

	ASCE 7-98	ASCE 7-95	ASCE 7-88
Averaging Time	three seconds	three seconds	Varies, with velocity (fastest mile wind speeds)
Importance Category	No interpolation	No interpolation	• Interpolated within 100 miles (161 km) of hurricane coastline
Gust Response Factors	• Rigid: 0.85 or • Flexible: calculated by formula	• Rigid: 0.80 – Exp. A & B, 0.85 – Exp. C & D • Flexible: calculated by a rational analysis incorporating dynamic properties of MFRS	• Gust effect changes with height. • Table for four exposure categories up to 500 ft (152 m)
Internal Pressure Coefficient	• Open = 0.00 • Partially-Enclosed = ± 0.55 • Enclosed = ± 0.18 • Total area of openings in the wall receiving positive external pressure exceeds area of openings in the balance of the building (± 0.80) AND • Total area of openings in a wall that receives a positive external pressure exceeding 4 ft^2 (0.3716 m^2)	• Open = 0.00 • Partially-Enclosed = -0.30 • Enclosed = ± 0.18 • Buildings in hurricane zone with basic wind speed ≥ 110 mph or in Hawaii AND • Glazed opening below 60 ft (18.3 m) are not designed to resist wind-borne debris or are not protected from debris input	• Fully-Enclosed = ± 0.25 • Buildings with one wall opening exceed sum of percentage of openings in remaining walls and roofs by 5% AND • Percentage of openings in any one of the remaining walls ≤ 20% + 0.75 – 0.25
Topographic Factor	Abrupt changes in general topography buildings sited on upper half of hills and escarpments: $K_{zt} = (1 + K_1K_2K_3)^2$ Not required if $H/L_h < 0.2$ Exp. C & D $H < 15$ ft (4.6 m) Exp. A & B $H < 60$ ft (18.3 m)	Abrupt changes in general topography buildings sited on upper half of hills and escarpments $K_{zt} = (1 + K_1K_2K_3)^2$ Not required if $H/Lh < 0.2$ Exp. D $H < 15$ ft (4.6 m) Exp. C $H < 30$ ft (9.1 m) Exp. A & B $H < 60$ ft (18.3 m)	None
Wind Directionality Factor	• 0.85 – used when applying load combinations per Sections 2.3 and 2.4 • Factored load strength design – 1.6 • Allowable stress design – 1.0	Included in wind load factor: • Factored load strength design – 1.3 • Allowable stress design – 1.0	Included in load factor

	ASCE 7-98	ASCE 7-95	ASCE 7-88
Wind Pressure Combinations for Full and Partial Loading for Buildings > 60 ft (18.3 m)	• Full pressure on projected areas ⊥ to the principal axis considered separately • 75% of load acting on 50% of projected area	• Full pressure on projected areas ⊥ to the principal axis considered separately • 75% of load acting on 50% of projected area	None
External Pressure Coefficients for Components and Cladding for H < 60 ft (18.3 m)	• Saw tooth, gable roofs, and walls • Gable/hip roofs, $10 < \theta < 30$ • Gables $30 < \theta < 45$ and $3 < \theta < 10$ • Stepped roofs • Multispan gable • Monoslope, $10 < \theta < 30$	• Gable roofs and walls • Hipped roofs • Stepped roofs • Multispan gable roofs • Monoslope and saw tooth roofs • Multispan saw tooth roofs	• Gable roof building roofs and walls

Design Procedures of ASCE 7-98

The first step to determine wind loads on a building component is to select the appropriate design wind speed from the basic wind speed map provided in ASCE design standards. Past versions of ASCE 7 wind load design standard supported two design procedures; an analytical method and the wind tunnel study. ASCE 7-98 has added a third procedure, Procedure 1, discussed below. In ASCE 7-98, the presentation of figures and tables has also been improved, and information is more clearly presented than in previous versions of ASCE 7. A summary of the three procedures follows:

Procedure 1: Simplified Procedure – This procedure is applicable to relatively-common, regular-shaped, low-rise, diaphragm (shear wall) buildings (roof height less than or equal to 30 ft (9 m)) and with roof slopes of less than 10°. The building must not be classified as a flexible building as specified in the commentary of the standard nor have expansion joints or separations. It must be located in an area that has no topographic effects.

Pressures for roof and wall loads can be selected directly from a table for the applicable basic wind speed. Values for components and cladding loads are provided for enclosed and partially-enclosed buildings. Values are tabulated for Exposure B, and multiplying factors are provided for Exposures C and D. The simplified procedure is not to be used for Exposure A because of greater uncertainty of wind load distribution.

Procedure 2: Analytical Procedure – This method calculates wind load using formulae provided in ASCE 7-98. In order to determine the design wind pressures on components and cladding, the designer determines the basic wind speed, directionality, importance and topographic factors, velocity pressure coefficients, and internal and external pressure coefficients. These values are included in a series of figures and tables in the standard.

Procedure 3: The Wind Tunnel – Wind tunnel tests are recommended when the building or structures have one or more of the following conditions: The structure is irregular in shape, flexible, subject to buffeting by the wake from upwind buildings, and/or subject to accelerated flow caused by local topographical effects or channeling. ASCE 7-98 limits the reduction that is allowed for shielding effects from upstream obstructions to 80% of the lowest wind loads calculated using the analytical procedure of ASCE 7 wind load provisions.

Sample Problems using ASCE Design Standards

The following sample problems demonstrate wind load calculations for exterior wall components and cladding using ASCE 7-98, 7-95, and 7-88 for a building in flat terrain and for the same building located on a prominent topographic feature. The wind pressures are determined for a wall component located in the edge zone and near the roof.

Sample Problem No. 1:

A building has the following dimensions: 40 ft (12.2 m) x 100 ft (30.5 m) in plan and with a mean roof height of 80 ft (24.4 m) high. The building is located in Boston, Massachusetts, in a suburban setting (Terrain Category B), and the Engineer has been asked to determine the wind loads on a window. The effective wind area used in this problem is 20 ft^2 (9.3 m^2) (Figure 2). The basic design wind speed specified in ASCE 7-98 is 105 mph (three-second gust).

For comparison purposes, this sample problem includes wind loads at three locations above ground – at 15 ft (4.6 m), 40 ft (12.2 m), and 80 ft (24.4 m) – and determines design wind pressures on the windward and leeward walls of the building. As is shown in the following tables, wind design load on the leeward wall differs in corner and field zones, and the pressures on the windward wall vary with height above ground.

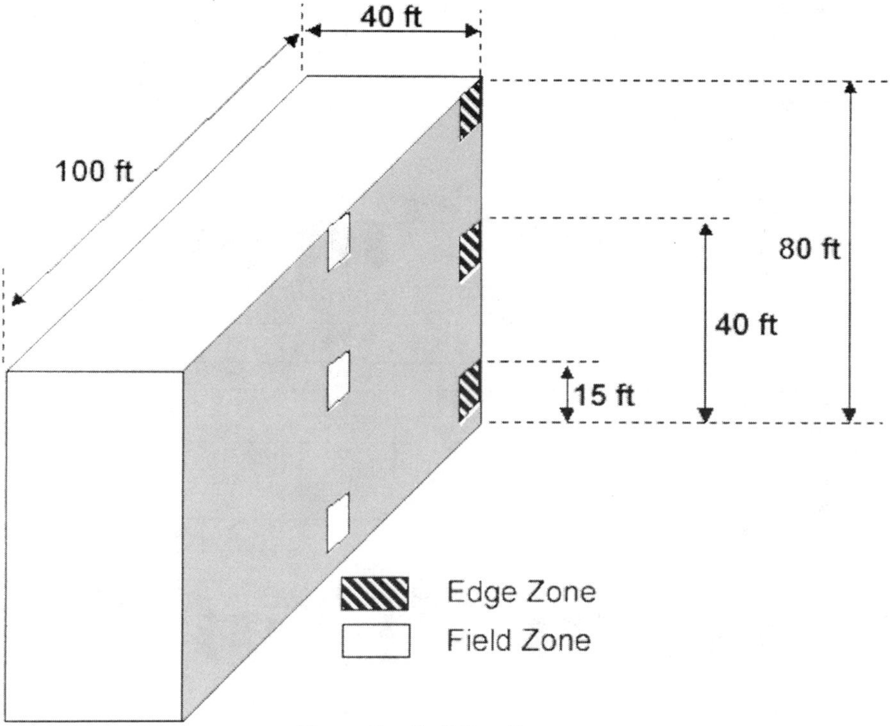

Figure 2 – *Building Geometry*

Table 3 – *Results of Maximum Design Wind Pressures (z = 80 ft (24.4 m)) for a 20 ft^2 (6.1 m^2) Effective Wind Area Cladding Element (All values are in psf. 1 psf = 0.04788 kN/m^2)*

ASCE Version	Velocity Pressure q at 30 ft (10 m)	Negative Pressure - Leeward Wall				Positive Pressure – Windward Wall	
		Enclosed Building		Partially-Enclosed Building			
		Corner	Field	Corner	Field	Enclosed Building	Partially Enclosed
7-98	22.3	-44.2	-24.1	-52.4	-29.0	+24.1	+32.4
7-95	28.8	-60.5	-34.6	-74.9	-49.0	+34.6	+49.0
7-88	15.7	-43.2	-19.6	-51.0	-32.4	+19.6	+27.5

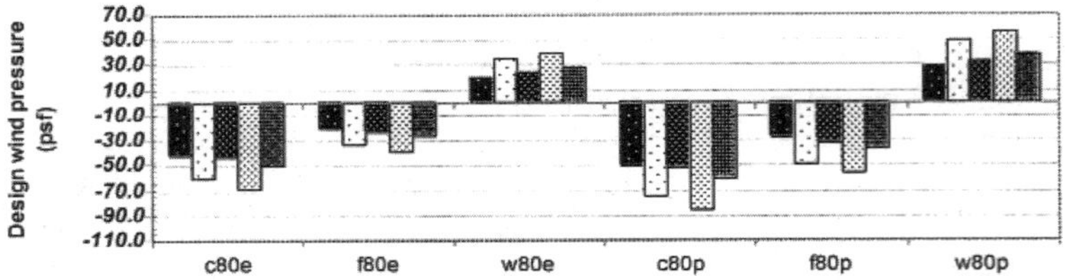

	c80e	f80e	w80e	c80p	f80p	w80p
ASCE 7-88	-43.2	-21.2	19.6	-51.0	-29.0	27.5
ASCE-7-95	-60.5	-34.6	34.6	-74.9	-49.0	49.0
ASCE 7-98	-44.2	-24.1	24.1	-52.4	-32.4	32.4
Kzt 7-95	-69.5	-39.7	39.7	-86.1	-56.3	56.3
Kzt 7-98	-50.8	-27.7	27.7	-60.2	-37.2	37.2

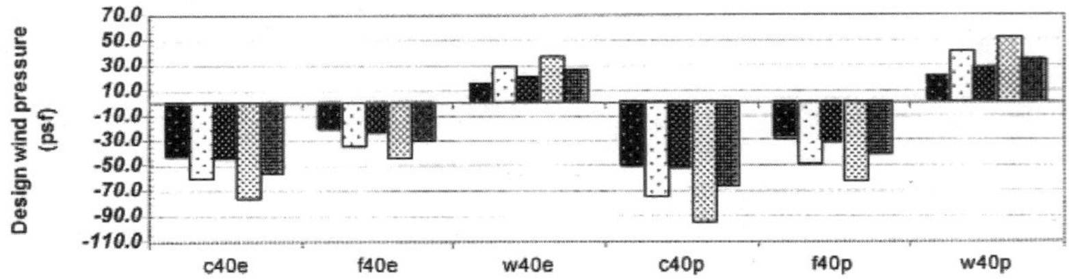

	c40e	f40e	w40e	c40p	f40p	w40p
ASCE 7-88	-43.2	-21.2	14.5	-51.0	-29.0	20.3
ASCE-7-95	-60.5	-34.6	28.3	-74.9	-49.0	40.0
ASCE 7-98	-44.2	-24.1	19.7	-52.4	-32.4	26.4
Kzt 7-95	-76.8	-43.9	35.9	-95.1	-62.2	50.8
Kzt 7-98	-56.1	-30.6	25.0	-66.6	-41.1	33.6

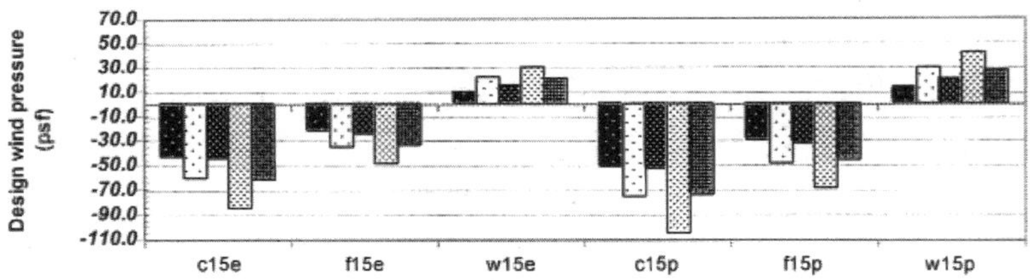

	c15e	f15e	w15e	c15p	f15p	w15p
ASCE 7-88	-43.2	-21.2	9.4	-51.0	-29.0	13.2
ASCE-7-95	-60.5	-34.6	21.2	-74.9	-49.0	30.0
ASCE 7-98	-44.2	-24.1	14.8	-52.4	-32.4	19.8
Kzt 7-95	-84.8	-48.4	29.7	-105.0	-68.6	42.1
Kzt 7-98	-61.9	-33.8	20.7	-73.5	-45.3	27.8

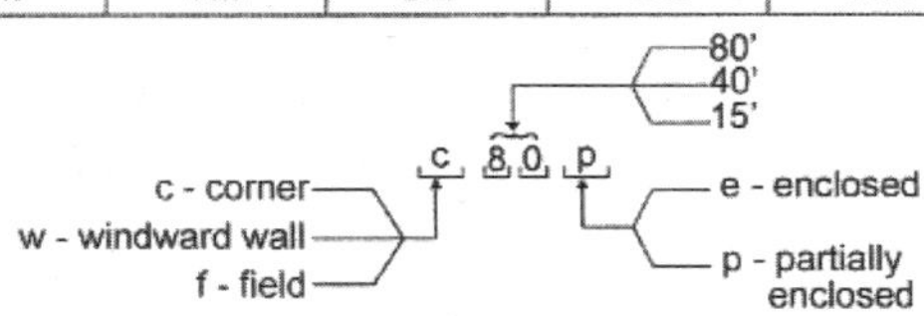

Figure 3 – *Summary Results of Wind Design Load Comparisons per ASCE 7 for a 20 ft^2 (6.1 m^2) Opening*

For the case of a partially enclosed building, Table 3 shows there is a slight decrease in calculated maximum wind design loads using ASCE 7-98 versus ASCE 7-88 provisions for wall cladding in the field of the wall near the top of the building. Similarly, the maximum wind design load on wall cladding at the corners of this building would increase slightly. In contrast, using ASCE 7-95 provisions results in significant increases in maximum wind design loads; i.e., a 51% increase for components and cladding in the field of the wall and a 46% increase for wall components and cladding at the corners of this (partially-enclosed) building.

ASCE 7-98 wind design provisions also produced increases in maximum wind design pressures on the windward walls over the ASCE 7-88 provisions. The increase in positive pressure on the wall components and cladding with ASCE 7-98 (average about 20%) are much less than the 77% increase in wind design pressure on windward walls determined using ASCE 7-95 provisions.

Figure 3 presents a summary of all data, comparing wind design loads at leeward and windward walls for components at three wall heights: 15 ft (4.6 m), 40 ft (12.2 m), and 80 ft (24.4 m) for corner and field conditions and the effect of topography factors. These results show similar trends to an earlier study performed for wind loads on roofing by the Single Ply Roofing Institute [*6*].

Sample Problem No. 2:

K_{zt} – Topographic Factor for Velocity Pressure – The same building is now built on the rise of a two-dimensional escarpment. The building is sited 40 ft (12.2 m) away from the edge of a cliff having the upwind profile of a 200 ft (61 m) rise in an 800 ft (243.8 m) run (Figure 4). Find the design load at 15 ft (4.6 m), 40 ft (12.2 m), and 80 ft (24.4 m) above ground level for the cladding assuming a partially enclosed building (conservative case).

Topographic factors take into account wind speed-up over hills and escarpments. Buildings sited on the upper half of an isolated hill or escarpment can experience wind speeds that are significantly higher than buildings situated on level ground. The velocity pressure is calculated using a topographic factor, K_{zt}, which is determined by three multipliers, K_1, K_2, and K_3, shown in Table 4. K_{zt} is equal to unity for buildings sited in flat terrain.

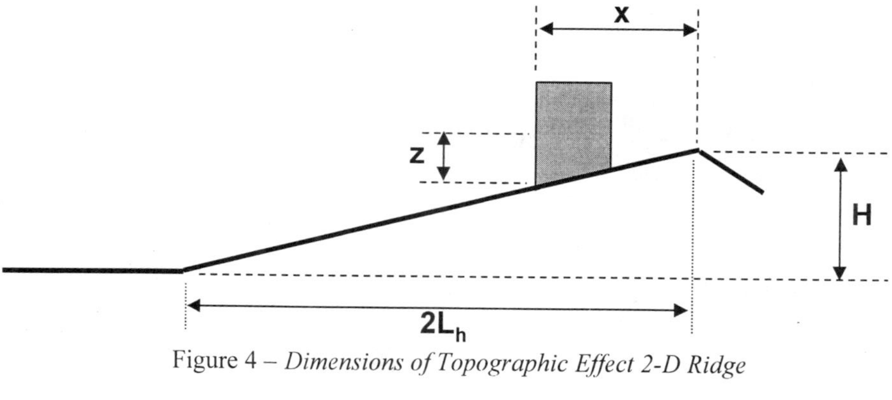

Figure 4 – *Dimensions of Topographic Effect 2-D Ridge*

Based on above topography, the multipliers are determined as follows:

$$H/Lh = 200 \text{ ft}/400 \text{ ft} = 0.5; K1 = 0.43 \quad (1)$$

$$x/Lh = 40 \text{ ft}/400 \text{ ft} = 0.10; K2 = 0.976 \quad (2)$$

$$z/Lh = 15 \text{ ft}/400 \text{ ft} = 0; K3 = 0.902 \quad (3)$$

$$K_{zt} = (1 + K_1K_2K_3)^2 \quad (4)$$

Table 4 – *Effect on Velocity Pressure for a Mid-Rise Building Located on a Ridge*

Height (ft)[2]	K_1	K_2	K_3	K_{zt}
15	0.72	0.975	0.90	1.40
40	0.72	0.975	0.74	1.27
80	0.72	0.975	0.55	1.15

[2] 1 ft = 0.3048 m.

The topographic factor produces a significant increase in wind design load on the components and cladding of this building, as indicated respectively by the 40%, 27%, and 15% increase factors at 15 ft (4.6 m), 40 ft (12.2 m), and 80 ft (24.4 m). The engineer is advised to exercise caution in using this factor, as wind speed-up factors should be used where buildings are sited on isolated features that are unobstructed upwind by similar topographic effects of comparable height, for 100 times the height of the topographic feature or 2 miles (3.22 km), whichever is less.

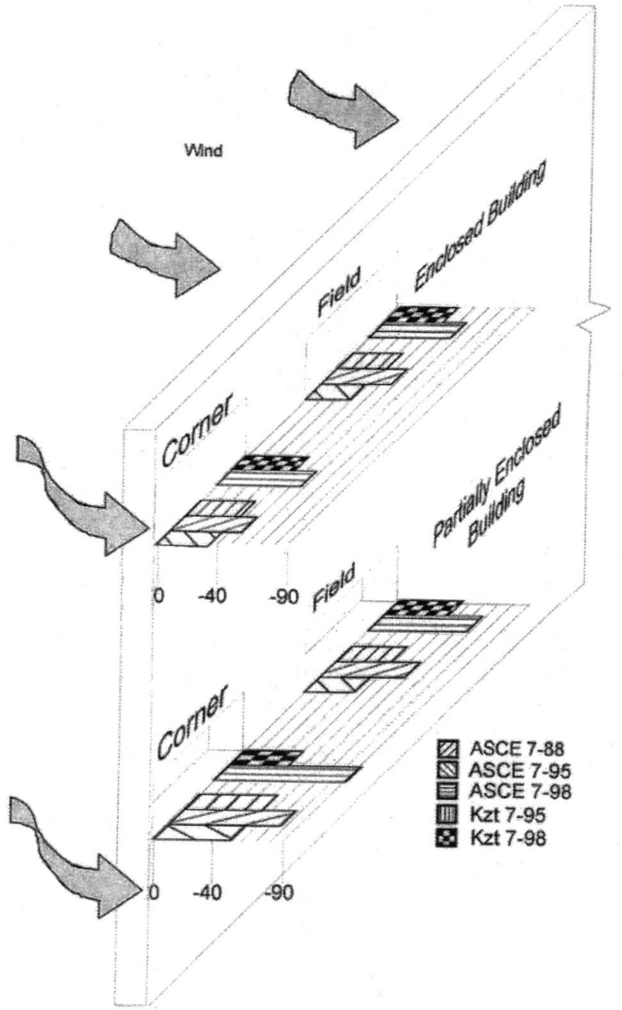

Figure 5 – *Comparing ASCE 7 Wind Design Loads on Leeward Wall for a Building on Flat Topography with one on a 2-D Ridge*

Sample Problem No. 3:

Wind Speed Changes in ASCE 7 Wind Contour Maps – Revisions to wind speed contour maps have occurred reflecting better data on wind speeds in interior areas of the country. The contour maps have benefited from additional wind speed measurements that showed that wind speeds away from coastal areas are not as severe as predicted by the contour map in ASCE 7-95. For example, the basic wind speed for Orlando, Florida, is reduced in ASCE 7-98 about 23% from the value in ASCE 7-95.

Table 5 – *Design Wind Speeds for Locations (mph)*[1]

Location	Local Code Design Wind Speed	ASCE 7-88 Fastest Mile Wind Speed	ASCE 7-88 Equivalent Three-Second Gust	ASCE 7-95	ASCE 7-98
Boston, MA	90 (FM)	85	103	110	105
Providence, RI	90 (FM)	90	108	125	110
Miami, FL		110	129	150	147
Springfield, MA	70 (FM)	80	97	90	100
Columbia, SC		75	92	90	95
Orlando, FL		95	113	137	105
Houston, TX		90	108	110	115

[1] 1 mph = 0.44704 m/s.
[2] FM = Fastest mile wind speed (mph).

Sample Problem No. 4:

Simplified Procedure Versus Analytical Procedure – Table 6 shows a comparison of wind loads obtained for components and cladding using ASCE 7-98's Procedures 1 and 2 for the same building located in a suburban exposure, Category B in Providence, Rhode Island.

For this example, assume the building dimensions are 40 ft (12.2 m) x 60 ft (18.3 m) in plan, with a low slope roof having a mean roof height of 33 ft (10 m). The pressures in the following table are for component and cladding with an effective wind area of 50 ft^2 (4.6 m^2). The building is a rigid diaphragm structure with no expansion joints or separations. The basic wind speed is again 110 mph (49.1 ms^{-1}), three-second gust.

Table 6 – *Comparison of Negative Design Pressures for Components and Cladding Loads Obtained Using the Simplified and Analytical Procedures of ASCE 7-98.*

Location	Simplified Procedure Enclosed Building (psf)[4]	Simplified Procedure Partially-enclosed Building (psf)	Analytical Procedure Enclosed Building (psf)	Analytical Procedure Partially-enclosed Building (psf)	Difference in wind pressure for enclosed building (simplified/analytical)
Corner	-25	-31	-24	-30	+4.2%
Field	-22	-28	-19	-26	+15.8%

[4] 1 psf = 0.04788 kN/m^2.

The design wind pressures obtained using the analytical procedure are consistently lower than the design pressures obtained by the simplified procedure, but in this case, the absolute differences in magnitude are not significant in absolute terms.

Structural Performance Tests of Exterior Wall Components and Cladding

Engineers and architects need to be aware of the application of new tests for components and cladding. As the number of available cladding systems increases and components become more complex to assemble, the potential for assembly defects and wind damage increases. In earlier building periods, which were dominated by heavy construction materials, the standardized testing of building assemblies was not an integral part of the building process as it is today. Yet there are numerous examples of buildings that survived the test of time. Modern construction requires numerous tests for exterior walls to aid the design process.

Historic Building Walls

From historic times up to the beginning of the twentieth century, the testing of wall components was not a major issue for exterior building walls because the self-supporting masonry wall systems possessed sufficient capacity to resist wind loads. Exterior walls were designed empirically, basing new systems on the successful performance of previous ones. A construction tradition developed in which the building walls had conservative safety factors to resist lateral loads. Exterior masonry walls were relatively massive, typically ranging from 12 in. (0.31 m) to 18 in. (0.46 m) thick in smaller buildings. In larger buildings, the wall dimensions increased to 2 ft (0.61 m) to 4 ft (1.2 m) or even larger. The window and door openings were small and framed with arches, acting in compression or with timber members set into the wall. Interior finishes were used sparingly, or they consisted of durable and breathable stuccos or other materials compatible with the masonry.

Because of their inherent strength and design, masonry walls supported their own weight, had relatively low compressive stresses, and had minimal water penetration problems. Thick walls acted as a reservoir for absorbed water, which, over time, was able to evaporate to the outside with minimal seepage to the interior. Walls constructed in this way had very long life expectancies (fifty to sixty years as a minimum). The construction of these walls was performed with few trades under the direction of a single builder, architect, or engineer, further minimizing the coordination issues the industry faces today.

Contemporary Exterior Wall Components and Cladding

Structural performance tests are necessary to improve the ability of components and cladding to resist high winds. As implied by the description of exterior building walls as "components and cladding," wall assemblies consist of many parts working together to form the whole system. In most cases, the components and cladding depend on an independent internal structural frame for support. The reason for these new changes is the drive to build lighter, stronger walls at reduced construction costs.

The design and construction of exterior walls is completely changed today from earlier times, and new methods are continually evolving. Exterior walls can be comprised of multiple systems, many by different manufacturers that may not always be compatible. The work now involves many trades and materials, extensive coordination requirements, and protection of fragile materials from damage during the construction process. Wall openings are larger to provide greater natural light into the structure. Cladding attachments must also be protected from corrosion and moisture damage. The following table compares the traditional and curtain wall design systems.

Table 7 – *The Traditional Masonry Wall and Modern Curtain Walls Comparison*

Traditional Wall	Modern Curtain Wall Components
• Solid masonry wall over 12 in. (0.31 m) thick • Metal flashing at roof and ground • Exposed masonry on interior or durable stucco finish	• Steel or concrete frame • Aluminum and glass windows, punched opening or in continuous strips • Masonry veneer, 4 in. (0.12 m) thick, tied to a backup • Metal flashing at roof, and ground • Metal flashing at window sills and headers and lintel beams • Metal fasteners between curtain wall and backup • Backup – steel and gypsum sheathing or wood stud • EIFS, stucco on steel frame, plywood – exterior waterproofing coating • Membrane waterproofing • Adhesive/ sealant • Interior finishes, wallpaper, paints

Reducing the mass and cross-sectional dimension of exterior walls reduced the gravity loads that the building frame must support, but the wind loads are essentially unchanged for similar-shaped buildings. In addition, using lighter materials connected to the main frames introduced different load paths and potential stress concentrations that did not exist before. The damage to cladding after hurricanes has been increasing, and the costs of damage and repairs have raised the need to improve the design of building walls.

Structural Performance Testing

Field studies after major wind events confirmed the poor performance of many cladding systems in high winds. Smith [7] reported that roof cladding systems are liable to fail below their design values, even for properly designed and installed systems. The same is true of wall cladding systems. Engineers use structural performance test results to find the ultimate failure loads of a wall assembly and to determine an acceptable factor of safety and allowable design load. The allowable load value is meaningful only if the following three conditions are met:

1) The test method used represents realistic loading on the component.
2) The tested specimen is constructed, as it would be in its installed location, using an equivalent standard of care and workmanship.
3) A rational determination of the safety factor is made taking into account the variability in material properties, manufacturing, and construction errors.

Suggested reasons for the continued failures of cladding systems include a limited understanding of their behaviors, and also the inappropriate test procedures used to determine wind load design capacity. In addition to the structural performance tests, building cladding tests determine the water penetration resistance and the overall durability of the materials exposed to long-term weather effects. For structural effects in exterior walls, two types of tests are necessary: material tests and structural performance tests, which predict how the wall system behaves under prescribed loads.

Structural performance tests provide the basis for engineering data used for design. Building codes, insurance companies, or consensus agreement within specific industries sometimes mandate which tests are necessary to certify product performance. In South

Florida, building officials have increased mandatory tests for building envelopes within hurricane-prone regions in the hope of reducing losses from the next hurricane landfall.

In wind uplift tests on metal roofing, researchers [*8*] found that cladding materials that undergo large deformations under pressure redistribute loads among fasteners, resulting in the overloading of some fasteners. The same effect is likely to occur with flexible wall cladding systems. Current performance tests do not recognize this potential problem and, indeed, provide a rating system for the assembly subjected to uniform static loads. While the actual loads on installed cladding may remain below the allowable design values, certain spatial load conditions may locally overstress a fastener and cause it to fail. In addition, for mechanically-attached cladding systems, such as metal sidings and mechanically-attached single-ply membranes, fasteners typically have little reserve capacity to resist overstressed conditions and fatigue, and the failure of a single fastener is sometimes sufficient to cause a zipper-like failure of the remaining fasteners along a row.

ASTM Structural Performance Tests

The American Society for Testing and Materials (ASTM) publishes many structural performance test protocols for curtain wall systems that have been adopted by building codes. The following three performance tests are becoming the basis for building wall wind design performance tests:

ASTM E330 – This test, first promulgated in 1967, determines the structural performance under the effects of wind loads on exterior windows, curtain walls, and doors under uniform static pressure. The test is applicable to curtain wall assemblies including but not limited to metal, glass, masonry, and stone components. ASTM E330 specifies ASCE 7 to determine the design wind loads, and it specifies a test load of 1.5 times the design load to be used in the test. The test protocol includes the following noteworthy comment:

> "Performance is a function of fabrication, installation and adjustment. In-service performance depends on the rigidity of the supporting construction, temperature and on the resistance of components to deterioration by other causes, including vibration, thermal movement and the authors add material degradation."

ASTM E1233 – was developed in 1988 along similar lines as E330 but to represent the long-term effects of repeated applications of wind loads or those loads that may be experienced in a hurricane or other extreme wind event. This test is designed for conditions when *load cycle effects are significant* and for unique constructions that have insufficient field performance data to establish performance criteria. The paper by LaTona et al [*9*] provides additional information on this test, as well as the benefits and uses for certain assemblies. E1233-00 provides three load criteria, a simplified life cycle test, a hurricane test spectrum, and an extreme wind test spectrum. Judgment must be exercised in determining if load cycle (or fatigue) effects are significant; however, the protocol recommends the test be used when the building envelope consists of any of the following:

- Metallic assemblies
- Threaded fasteners that are not self-locking
- Welded assemblies
- Assemblies with notch effects
- Masonry veneer on flexible backup
- New materials, composites, and brittle materials (EIFS, stuccos, plastics)

ASTM E1886 – Post-hurricane investigations found that a significant amount of damage to building envelopes is caused by windborne debris. ASTM 1886 provides a method to determine the ability of wall cladding elements to remain unbreached during a hurricane. The two-fold test procedure involves initial missile impacts on three specimens of the cladding element with large and/or small missiles, followed by a cyclic pressure test.

Windborne debris typically consists of framing lumber, roof tiles, and sheet metals (represented by 2 in. (0.05 m) x 4 in. (0.1 m) lumber between 4.5 lbs (2.0 kg) and 9 lbs (4.1 kg)) traveling between 40 and 80 fps (or 12.2 and 24.4 m/s). The small missiles used in the test represent roof gravel. The large missile size and impact speed is selected depending on the building location within hurricane wind speed zones. The design wind load is determined using ASCE 7.

Table 8 – *Summary of ASTM E1886 Impact and Cyclic Test Procedures*

Test	Description		
	The large missile test required for openings below 30 ft (9.1 m) above ground. A small or large missile test required for openings more than 30 ft (9.1m) above ground.		
Large Missile Test	No. 2 or better Southern Yellow Pine/Douglas Fir 2 x 4 lumber between 4.5 lbs (2.0 kg) and 9 lbs (4.1 kg). Travel speed between 40 fps (12.2 m/s) and 80 fps (24.2 m/s).		
Small Missile Test	Twenty spherical steel balls with diameters 8 mm (0.1 ft), weighing 2 g (0.004 lbs), and traveling at a speed of 130 fps (39.65 m/s)		
Cyclic Pressure Test	Loading Sequence	Sequence Range	Number of Cycles
	1	Positive 0.2P to 0.5P[5]	3,500
	2	Positive 0.0P to 0.6P	300
	3	Positive 0.5P to 0.8P	600
	4	Positive 0.3P to 1.0P	100
	5	Negative 0.3P to 1.0P	50
	6	Negative 0.5P to 0.8P	1,050
	7	Negative 0.0P to 0.6P	50
	8	Negative 0.2P to 0.5P	3,350
Acceptance Criteria	No opening more than 5 in. (0.13 m) long or large enough to allow a 3 in. (0.08 m) sphere to pass (non-porous system)		

[5] P denotes the maximum inward (positive) and outward (negative) air pressure differentials defined by the publishing authority.

Other Structural Performance Tests

Many organizations produce and conduct structural performance tests for exterior cladding systems, including ASTM, Underwriters' Laboratories, Factory Mutual, Metal Building Manufacturer's Association, National Roofing Contractors Association, APA – The Engineered Wood Association, and the Single-Ply Roofing Institute. The drive for new and improved testing is to improve performance, reduce insurance liability, and provide safer buildings. The following chart provides a partial list of structural performance tests used for components and cladding:

Table 9 – *Testing Protocols for Components and Cladding (all tests without designations are from ASTM)*

Material	Static Wind Loads	Impact and Dynamic Wind Loads	Water Penetration	Weather Durability	Thermal Loads
Masonry Veneer	E330	E1233	E331 Rilem Test		
EIFS and Stucco	E330	E1886 E1233			
Metal Siding	E1592 E330	E1886 E1233			
Plastic Skylights		E1886 E1233	E331		NFRC 100
Plastic Panels	D5206 E997				
Glass	E997 E998 E1300		N/A		
Glass and metal framed windows	E330	E1886 E1233	E331, E547 E283 (air)		AAMA 1503 NFRC 100
Clay & concrete Tiles	SSTD 11-97			C1492	
Membrane Roofing	FM4471 UL 1897	UL580			
Metal Roofing	FM 4470 E1592		E1646		
Metal Flashing	ANSI/SPRI ES-1 1998				
Stone	C1201				
Sealants				AAMA 850	
Fasteners	ANSI/SPRI FX-1-1996				

Many components of the building envelope are currently tested to determine their resistance to wind loads. Tests have been developed for both roofing and wall elements and dynamic test protocols are sometimes used. The building envelope professional should be familiar with these tests and their intent for glazed openings, curtain wall systems, metal siding, and metal roof edge flashing, among other materials. Very few materials have tests that actually represent the wind's load effects on cladding in ways that duplicate the spatial changes in wind loads as well as the time component. The effect of spatial load variation has not been the subject of previous experimental work. However, a theoretical study of load distribution performed with a finite element analysis on mechanically-attached single-ply roof membranes demonstrated that most of the peak loads near to a fastener are transferred directly to that fastener with little distribution.

In 1994, Broward and Dade Counties in Florida implemented a new building code, which includes provisions that wall elements (including windows) for all buildings must pass county-approved missile impact and wind loading tests before buildings will be granted certificates of occupancy. In 1995, Palm Beach County, Florida, adopted similar testing procedures for glazed openings. The International Building Code, ICC 2000, also requires structural performance tests for glazed openings.

Metal Flashing – The Single-Ply Roofing Institute (SPRI) has developed a design guide and test standard for metal edge flashing, ANSI/SPRI ES-1-98, "Wind Design Standard for Edge Systems Used with Low-Slope Roofing Systems." This standard defines design loads and testing methods and uses static loads hung from a metal flashing mockup to obtain ultimate static failure loads.

Roofing – This paper deals with building walls, but the roof, as part of the overall envelope, must also be designed for wind load resistance. Previous experimental research for flexible roofing systems compared the uplift tests by Factory Mutual and Underwriters' Laboratories, which provide static and pseudo-static guidelines. A comparison of these tests can be found in Prevatt [*10*]. Further work on roofing includes a new dynamic test procedure [*11*] developed by a consortium of industry researchers and manufacturers and led by the national Research Council, Canada, which provides a test method for dynamic uplift wind resistance of mechanically-attached single-ply membranes.

Masonry – Not always considered as susceptible to wind or impact loading, but wind forces should be considered in designs especially for renovation and repairs to existing buildings. Test specimen weight and projectile speed used in ASTM 1886 was chosen because the projectile was shown to penetrate masonry. The assessment of allowable loads is determined by the flexural strength of the wall.

Precautions When Interpreting Performance Test Results

Some cladding materials lack structural performance tests, and test data from one material is not easily comparable to others. For example, if tests of two materials obtained the same ultimate failure pressure for a metal panel system and an aluminum- and glass-framed window system, the results may not establish the same wind design loads for both systems because the safety factors for each material are likely to be different. The same is true for different cladding systems, depending on methods of attachment and substrate materials.

In addition, although performance tests can be used for many cladding components, the relationship of the results between different systems must be determined with caution. For example, the factor of safety of the masonry components is likely not going to be the same as that for EIFS or plastic or metal siding. The structural load paths attaching the different materials also will affect the interpretation of results.

Some tests provide the ultimate capacity of the single component, neglecting possible serviceability problems due to incompatible performance at the edges between dissimilar materials. The pullout strength of a metal siding fastener may provide wind resistance, but what is also important to the engineer's design is the effect of thermal movement or fatigue loading around the fasteners that creates holes for water penetration. Low-level fluctuating loads may not cause catastrophic failure but may initiate hidden damage that, when left unchecked, will lead to additional failures and weaken the system before the major event.

Conclusions

The building envelope industry is a constantly evolving industry that must respond to changes and advancements in material building science and technology. With the vast numbers of systems to choose from, building envelope professionals need to keep pace with many emerging cladding technologies and construction techniques and to service the construction market. The construction of modern wall systems involves dozens of trades, materials, installation techniques, and manufacturers that the designer must be familiar with. These issues have caused numerous performance tests to become available to the designer but, with little coordination of their use, the numbers of tests have multiplied, and the overall effect on wall performance is not known and is difficult to quantify. There is an increased potential for design errors occurring.

Except for fully-engineered curtain wall systems and new buildings constructed in Metro-Dade and Broward Counties, wind-resistant structural performance tests are not universally nor systematically applied. The jurisdictions that require structural tests only on windows and doors and disregard the remainder of the building envelope can expect wind damage to still occur when flying debris penetrates wall components, or components suffer fatigue failure at their attachments. The jurisdictions that require structural performance tests for all building envelope components will see the largest improvement in the wind resisting performance of buildings.

The testing of individual windows for installation in punched openings has benefited from extensive proof testing, research, and development to minimize impact damage. Consumers beyond the Florida region can now specify impact-resistant windows and doors, and they have a wider choice of tested systems to choose from. Current design criteria still do not cover all conditions, and designers are warned that sometimes, the minimum requirements specified in building codes may not be enough. Professional judgment is required for every specific case.

Recommendations

The author makes the following recommendations:

- ASCE 7-98 wind design load provisions reflect current knowledge of how extreme wind affects buildings and structures. The ASCE 7-98 wind design values are slightly higher than those predicted using ASCE 7-88, and they are significantly less than design loads predicted by ASCE 7-95. Cladding designers should consider using the ASCE 7-98 wind load provisions instead of relying on ASCE 7-95 or ASCE 7-88.
- Isolated topographic effects can increase wind design loads on buildings constructed on hills and escarpments from 20% to more than 100%. However, designers should exercise caution when determining the applicability of this factor.
- Structural performance tests for building curtain walls, such as ASTM 330, 1233, and 1886, provide a starting point for the engineering design of cladding systems. The designer is responsible for selecting appropriate tests for every given curtain wall or cladding application. Published test results will not eliminate the need for thorough engineering analysis. Design professionals need to become familiar with the interpretation, use, and potential misuse of published performance test data.

- Long-term exposure to wind loads affect cladding performance, and research and laboratory testing is needed to calibrate the loss in cyclic performance with the ultimate load capacity given by a single laboratory test. The risk of fatigue failure should be considered for new materials especially with flexible, lightweight systems.
- Scientific procedures are needed to predict structural performance, material durability, and service life of cladding materials that significantly degrade over time. For materials where significant degradation can be expected, the material's wind resistance may be reduced from tested results before the next design event.
- The cladding industry needs to develop means to systematically collect and analyze data on wind-induced damage to components and cladding for all exterior wall materials. Actual performance of cladding materials in actual wind events is the surest way to understand how similar systems will fail in future events.

Dedication: This paper is dedicated to the memory of the late Professor Dale C. Perry of Texas A&M University, a leader of the wind engineering community. Dale provided exceptional guidance, inspiration, and insight to me throughout my graduate studies. This paper grew out of our many conversations about wind/structure interaction and the future of structural performance testing of building envelope systems.

References

[1] Sparks, P. R., "Private communication with author," 8 March 2001.

[2] American Society of Civil Engineers, "ANSI/ASCE 7-95 Minimum Design Loads for Buildings and Other Structures," *American Society of Civil Engineers,* December 1995.

[3] American Society of Civil Engineers, "ASCE 7-98 Minimum Design Loads for Buildings and Other Structures," *American Society of Civil Engineers,* January 2000.

[4] American Society of Civil Engineers, "ANSI/ASCE 7-88 Minimum Design Loads for Buildings and Other Structures," *American Society of Civil Engineers,* July 1990.

[5] Smith, T. L. "Winds of Change: ASCE 7's New Wind Load Provisions," *Professional Roofing,* National Roofing Contractors Association, Rosemont, IL, April 1996.

[6] SPRI, "Wind Guides: Your Results May Vary," *RSI Magazine,* July 1999.

[7] Smith, T. L., "High-Wind Performance of Mechanically-Attached Single-Ply Roof Systems," *Professional Roofing,* National Roofing Contractors Association, Rosemont, IL, February 1996.

[8] Prevatt, D. O., Schiff, S. D., and Sparks, P. R., "A Technique to Assess Wind Uplift Performance of Standing Seam Metal Roofs," *Proceedings of the 11th Conference on Roofing Technology,* National Roofing Contractors Association, Rosemont, IL, September 1995.

[9] LaTona, R. W., Schwartz, T. A., and Bell, G. R., "New Standard Permits More Realistic Curtain Wall Testing," *Building Design & Construction,* October 1988.

[10] Prevatt, D. O., "Wind Uplift Behavior of Mechanically-Attached Single-Ply Roof Membrane Systems," Ph.D. Dissertation, Clemson University, Clemson, SC, May 1998.

[11] Baskaran, A. and Nabhan, F., "Standard Test Method for the Dynamic Wind Uplift Resistance of Mechanically Attached Roofing Systems," *Internal Report No. IRC-IR – 699,* National Research Council Canada, Ottawa Canada, September 2000.

Colin J. Williams,[1] *Greg J. Conley,*[1] *and John Kilpatrick*[1]

The Use of Wind Tunnels to Assist in Cladding Design for Buildings

---

Reference: Williams, C. J., Conley, G. J., and Kilpatrick, J., **"The Use of Wind Tunnels to Assist in Cladding Design for Buildings,"** *Performance of Exterior Building Walls, ASTM STP 1422*, P. G. Johnson, Ed., ASTM International, West Conshohocken, PA, 2003.

Abstract: Wind loads on a building are sensitive to a number of factors, including the wind speed approaching the site, building height and shape, and the local influence of nearby buildings on the wind flow patterns. Building codes attempt to allow for these factors by providing simple formulae for calculating design wind loads that will be at least conservative. The American Society of Civil Engineers (ASCE) 7 Standard [1] and most other building codes recognize that for irregularly shaped buildings or structures that may have unusual response characteristics it is advisable to undertake detailed wind load studies or use wind tunnel methods of analysis. Wind tunnel methods determine the wind loading on a structure with increased precision, which leads to more economical and risk consistent structural designs than do code calculation methods. This paper describes the wind tunnel method of determining cladding wind loads, and provides comparisons between the wind tunnel method and code calculation methods for a 22-story building.

Keywords: wind tunnels, cladding design, building geometry, exposure category, wind climate, pressure measurements, shielding

Introduction

One of the intents of the wind loading provisions in building codes and standards such as the ASCE 7 Standard is to provide minimum design wind loads that buildings and other structures should be designed to. Wind loads are sensitive to a number of factors, including wind speed and wind turbulence approaching the site, the building height and geometry, and the influence of nearby buildings on the local wind flow patterns. To specify design wind loads precisely for every possible building shape and surrounding environment would result in load provisions so complex as to be of limited

---

[1] Principal, Senior Engineer/Associate, and Senior Engineer, respectively, Rowan Williams Davies & Irwin Inc., 650 Woodlawn Road West, Guelph, Ontario, N1K 1B8

use to practicing structural designers. Therefore the writing of good codes and standards involves some compromise.

The approach taken by most codes and standards in predicting cladding wind loads on buildings, including the ASCE 7 Standard, is to provide simple formulae that include a measure of conservatism, as might be expected based on the approach taken in deriving the formulae. For small projects (e.g., less than 10 stories) with fairly simple geometries, the code formulae are probably of sufficient accuracy for design purposes and conservative results may not have a major cost impact. However, the ASCE 7 Standard recognizes that for structures with more complex geometry it is better to undertake detailed studies using wind tunnel tests since they yield more precise definitions of the design loads, and more economical and risk consistent structural designs than the code calculation method. Since most of the generic wind load provisions given in codes and standards are based on wind tunnel results, if the extra cost of the wind tunnel studies can be justified, it makes sense to utilize wind tunnels to predict the loads with precision.

Factors Influencing the Decision to Wind Tunnel Test

Building Geometry

Generally speaking, a building whose geometry deviates substantially from the simple shapes presented in the code provisions may be a candidate for wind tunnel testing. Irregularities in the building facade may include re-entrant corners, sudden changes in the facade such as setbacks, and architectural features such as fins or canopies. The formation of a vortex (a whirling region of separated flow) in re-entrant corners can significantly increase the negative wind pressures, or suctions, acting on the building facade, and cannot be predicted using code procedures. These shape dependant factors can be simulated directly in the wind tunnel.

Buildings that have open balconies or other similar details near the corners may experience significant local reductions in pressure [2]. The reductions in pressure at the corners of the building are due to the open balconies disturbing the wind flow as it comes around the corner of the building. Currently, there are no design procedures in the ASCE Standard to determine wind pressures at building corners that include the effect of open balconies on the pressure distribution. Understandably, code predictions of the wind pressures are conservative in this situation. Wind tunnel tests are necessary to measure the reduction, if any, in wind loading for this case.

Exposure Category and Shielding

Wind tunnel studies may be warranted for buildings with unusual exposures. For example, structures whose surroundings include unusual topography or multiple exposure categories may be prime candidates for testing. Generally, the exposure category that reflects the terrain characteristics of the site for each wind direction should be considered for design. However, code provisions usually require that cladding elements be designed using an open exposure (Exposure C - ASCE 7) unless terrain representative of other exposures prevails in the upwind direction for a specified distance. In some situations this will lead to conservative design wind loads, for example when a building is sited in a transition zone between categories defined in the code. An example of such a transition zone would be a building sited in a suburban fetch of terrain (Exposure B) extending between 5 and 10 building heights upstream, with open terrain (Exposure C) beyond. The code would require this building to be designed for Exposure C.

Similarly, in most codes, shielding of a building by its surroundings may not be considered for the calculation of cladding design wind loads. In many cases, the impact of the surroundings on a building is to reduce the wind loads acting on the cladding elements. However, this is not always the case [3]. Wind tunnel tests allow for variation in terrain categories and the effects of shielding to be simulated directly at model scale, and remove the conservatism introduced by the requirements of the code. Thus, the cladding is designed for a consistent level of risk with respect to the wind loads, which may result in substantial savings in the cost of the cladding for the structure.

Impact on Surrounding Structures

Many new developments in urban areas are required to demonstrate that the impact of a proposed structure on its surroundings is negligible, or at the very least, within limits covered by the use of appropriate load factors. New development may impact pedestrian level winds around existing structures or the distribution of wind pressures on neighbouring structures. These effects cannot be properly estimated by code procedures, and require the use of wind tunnel methods. For example, observations made during wind tunnel tests at the concept design phase may lead to changes, such as landscaping, wind fences, or canopies, to avoid impacts on neighbouring structures. The use of the wind tunnel early during the concept design phase may help to avoid costly litigation at later stages of development due to negative impacts on surroundings.

Another issue to be considered by a cladding consultant is the potential impact of changes in surroundings on the building they are designing. Wind tunnel tests frequently indicate wind pressures that are lower than predicted using code procedures. However, a change in surroundings, such as future developments, may increase wind loads on the study building. This may lead to overloads, should the cladding have been designed for the pressures measured in the absence of the future development. Recent studies have found that high overload situations caused by future changes in surroundings can be largely avoided by imposing a minimum value or "lower-cutoff" on the recommended wind loads from wind tunnel tests [4]. The ASCE 7-98 Standard recommends a minimum design wind pressure equal to the greater of i) 80% of the central zone wall pressures calculated using the ASCE 7 Standard or ii) the pressures predicted using wind tunnel methods.

Size of Building

It is generally recognized that wind tunnel methods predict wind pressure distributions on buildings more precisely than code analytical methods. Wind tunnel tests permit the measurement of exact magnitudes of the pressures on the building and their spatial distribution. Often, wind tunnel predictions may be lower than the code estimates. Unfortunately there is no specific guideline as to when a building should or should not be wind tunnel tested to realize cost savings in the cladding. Wind tunnel tests have been conducted on buildings ranging in height from only several stories to the tallest towers in the world. However, the majority of wind tunnel tests are performed on buildings higher than 10 stories in height. For low-rise buildings, designers of cladding and components must determine whether or not a cost benefit exists for using wind tunnel tests. Regardless of building size, wind tunnel tests are advisable when the building is of unusual shape or is situated in complex topography or surroundings.

Wind Tunnel Tests

Wind Tunnel Test Conditions

For wind tunnel methods to be permitted as the basis for design and the results accepted by code officials, the ASCE recommends that wind tunnel tests typically should meet all of the following conditions [1]:

1. The natural atmospheric boundary layer has been modeled to account for the variation of wind speed with height.
2. Relevant macro (integral) length and micro length scales of the longitudinal component of atmospheric turbulence are modelled to approximately the same scale as that used to model the building or structure.
3. The modeled building or other structure and surrounding structures and topography are geometrically similar to their full-scale counterparts, except that, for low-rise buildings meeting the requirements of 6.5.1[2], tests shall be permitted for the modeled building in a single exposure site as defined in 6.5.6.1.
4. The projected area of the modeled building or other structure and surroundings is less than 8 percent of the test section cross-sectional area unless correction is made for blockage.
5. The longitudinal pressure gradient in the wind tunnel test section is accounted for.
6. Reynolds number[3] effects on pressures and forces are minimized.
7. Response characteristics of the wind tunnel instrumentation are consistent with the required measurements.

Having satisfied these requirements, the data recorded during the wind tunnel tests should be representative of those expected at full-scale and is acceptable as an analysis tool by model building code organizations.

Wind Tunnel Test Model

The study building is usually constructed of acrylic material at a scale in the range of 1:400 to 1:500. Surrounding buildings are modeled within a radius of 500 m to 600 m. Beyond this radius, the wind tunnel facility simulates the upwind terrain by using a long working section with a roughened floor and specially designed turbulence generators at the upwind end (Fig. 1). This working section will generate the proper

[2] Clause in ASCE 7-98 outlining requirements for buildings whose design wind loads are calculated using code analytical procedures.

[3] A non-dimensional number that defines the ratio of inertial to viscous fluid forces.

variation of the mean wind speed and turbulence intensity profiles with height for the wind approaching the modeled surroundings [5]. As the approaching flow interacts with the modeled surroundings, wind flow patterns and turbulence levels similar to those expected at the site are simulated. The study building may be rotated in the wind tunnel through 360E to allow measurement of wind pressures for any wind direction, though typically 36 directions are recorded at 10E increments.

Figure 1- *Boundary Layer Wind Tunnel*

A typical wind pressure model may be instrumented with 250 to 1000 pressure sensors, or taps. Each pressure tap is connected via tubing (Fig. 2) to a pressure transducer. Pressure transducers are used to convert the measured pressure at the tap to an electrical signal collected by a computer for later analysis. Depending on the wind tunnel facility, pressure measurements at each tap are usually recorded for a time of 20 to 40 seconds at a sampling frequency of 500 to 700 times per second. Based on the design wind speed and length scale of the model study, this is approximately equal to a one hour sample at full scale at a sampling frequency of about 3 to 5 times per second. These sampling parameters are sufficient for determining the mean wind load acting on a cladding element, as well as the fluctuating components of the wind load due to the turbulence or gustiness of the wind.

The pressure taps are usually more highly concentrated in zones of the façade, which are anticipated to have higher wind pressures or rapidly changing pressure gradients. However, the wind-engineering practitioner may often query the client regarding the location of pressure taps; locations for which the client feels wind load information will be important for design can be identified prior to testing.

Many wind tunnel facilities have the capability to measure a large number of taps simultaneously and record time histories of the measured pressures. This allows for the summation or area-averaging of loads to be computed either on-line during a test, or if data storage capacity is sufficient, after the tests have been completed. This gives much more flexibility to measure a variety of load effects on features such as on large expanses of glass (such as atrium walls) or skylights [6].

Figure 2 - *Wind Tunnel Study Model*

The raw output from cladding pressure tests will usually consist of the mean, standard deviation (often referred to as RMS or root-mean-square), and peak positive and peak negative values of the pressure coefficients at each pressure tap. The data are collected for 36 wind directions at 10-degree intervals, which is typically sufficient to fully describe the dependence of the wind loads on wind direction. A pressure coefficient C_p is defined as

$$C_p = \frac{p - p_{ref}}{q_{ref}} \qquad (1)$$

where p is the local pressure on the model or building surface, and p_{ref} and q_{ref} are respectively the reference static pressure and reference dynamic pressure of the wind well above the site at the top of the simulated atmospheric boundary layer. The symbol p is used here to represent whichever particular pressure quantity we are concerned with, typically either a peak positive or peak negative value. Peak pressure coefficients may be estimated directly, but it is preferable to sample a population of measured peak values and then use statistical methods to evaluate the expected peak value during a storm [7]. Figure 3 illustrates a typical example of the variation of the mean and peak maximum and minimum pressure coefficients with wind direction for a location on a building wall. The value of the peak minimum pressure coefficient at this tap location is particularly sensitive to winds from the South-southwest (this may be significant if this corresponds with the direction of the highest probability of extreme winds). Note that wind directions of 0E, 90E, 180E, 270E and 360E in Fig. 3 correspond to winds approaching from the North, East, South, West and North directions respectively.

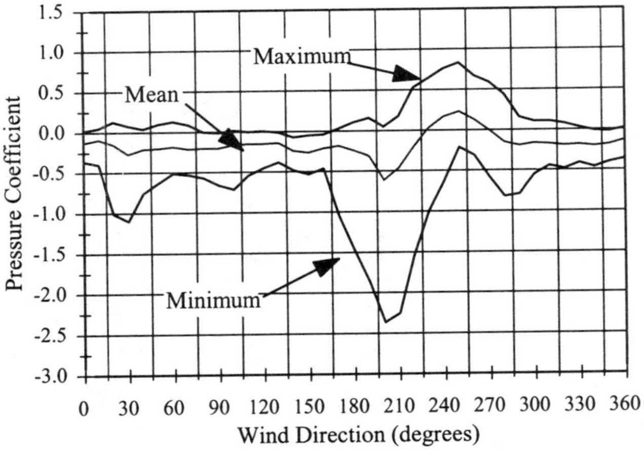

Figure 3 - *Pressure Coefficient Data from a Pressure Tap*

Determining Cladding Design Wind Loads

Before cladding design wind pressures can be derived from the raw wind tunnel data, due consideration must be given to the strength of the local wind climate. In order to predict the full-scale wind pressures from the wind tunnel data, these data must be combined with a statistical description of the local wind climate, which should consider the strength and directionality of the local winds. A method for combining raw wind tunnel data with the effects of the wind climate (known as the Up-crossing technique) have been discussed elsewhere [8,9]. To be consistent with American building code procedures, exterior cladding design wind loads are determined from the raw wind tunnel data using the 50-year return period wind speed derived from the wind climate model multiplied by an importance factor that varies depending on building classification. For buildings and structures that pose a threat to human life in the event of a failure or structures deemed essential facilities, the importance factor in effect implies cladding design wind loads equivalent to 100-year return period values.

Consideration must also be given to internal pressure. The design of cladding elements must consider wind loads on both the external surface of the structure (measured directly on the wind tunnel model) and loads on the internal surface of the cladding. The internal pressure can be strongly influenced by any openings in the building envelope which can be caused by open windows or screen doors, or by breakage due to flying debris in a storm. When the risk of openings during strong wind events is low (such as when high-impact glazing is specified or if the structure has no operable windows) internal pressure may be estimated satisfactorily using building code provisions for uniform leakage. When the risk of there being an opening is significant, the resulting effects on internal pressure can be incorporated into the determination of design loads in a rational way using the modified Up-crossing approach [10].

For final cladding design wind loads, the net wind loads (the difference between the external and internal wind pressures) are obtained by combining the external and internal wind pressures in a manner that will cause the worst load effect. For instance, the peak negative cladding design wind pressure is found by subtracting the positive internal pressure from the peak negative external pressure determined from the wind tunnel. Likewise, the peak positive cladding design wind pressure is found by subtracting the negative internal pressure from the peak positive external pressure determined from the wind tunnel.

Other considerations, such as the presence or absence of hurricane shutters, can influence cladding design wind pressures. These considerations should be discussed with the cladding consultant during the concept design and wind tunnel test phases.

Benefits of Wind Tunnel Testing - A Case Study

A recent consulting project at Rowan Williams Davies & Irwin Inc. (RWDI) involved wind tunnel testing of a 22-story office building in Phoenix, Arizona [11] to determine design wind loads for the cladding elements. The building is situated in a suburban terrain that includes many low buildings upwind of the study area with the downtown core immediately West and North of the study site (Fig. 4).

Figure 4 – *Phoenix, Arizona*

The study building is exposed primarily to suburban terrain (Exposure B - ASCE 7-98), except for sheltered wind directions ranging clockwise from the West through to the Northeast. Approximately 500 pressure taps were installed on the model. The plan shape of the building is elliptical with squared ends.

As noted earlier in this paper, building code analytical procedures would not allow for any reduction in wind loads due to the sheltering effect of the nearby surroundings for this building. However, the effect of shelter from surrounding buildings, namely a reduction in the design wind loads, was accounted for directly in the measured wind tunnel data. In order to predict the full-scale wind pressures acting on the building, the wind tunnel data was combined with a statistical model of the local wind climate. The statistical wind climate model used to determine the predicted peak pressures was based on surface wind measurements taken at Phoenix Sky Harbor International Airport. The raw surface wind measurements were adjusted to account for factors such as variation in the measurement height above ground over time and also the influence of surrounding terrain. The raw data for each wind direction were fitted with a Weibull[4] probability distribution to determine the probability of exceeding various mean hourly wind speeds from within each of 36 wind sectors at gradient height (Fig. 5). The fitted data are smoothed during subsequent analysis to represent a continuous wind speed probability distribution function.

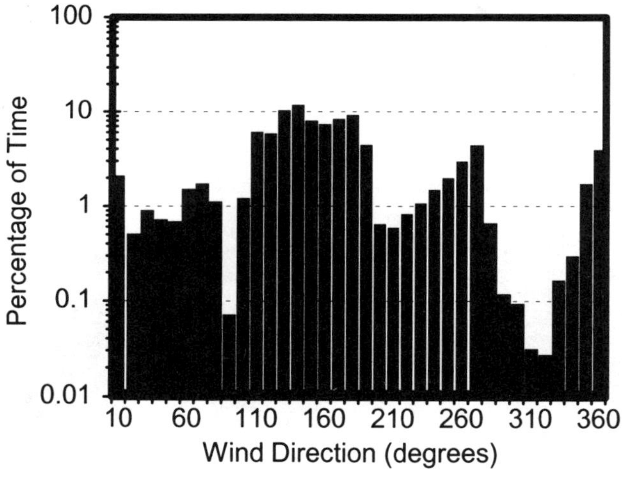

Figure 5 - *Raw Probability Distribution of 50-Year Design Winds - Phoenix, Arizona*

The design wind speed for the study building corresponds to a 50-year return period 3-second gust wind speed of 40 m/s at 10 m above ground in open terrain, which is consistent with the ASCE 7 Standard. Net cladding design wind pressures were derived from the external wind pressure distribution predicted from the wind tunnel study and internal wind pressures were estimated using a method which incorporates the internal compartmentalization of the building and probability of openings in the building envelope [10]. The 50-year cladding design wind pressures are shown in Fig. 6, zoned

[4] A two parameter mathematical model used to describe the probability of exceeding a particular magnitude of a random variable.

into 0.5 kPa (10 psf) increments. Feedback from cladding consultants suggests that this presentation of the design wind pressures is most useful during the cladding bid process.

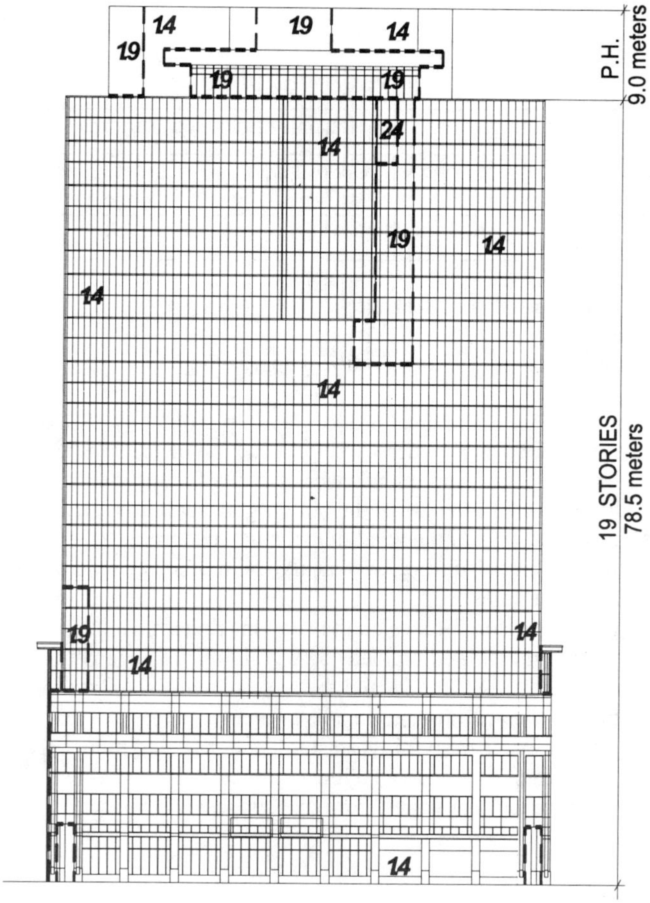

Figure 6 - *Recommended 50-Year Cladding Design Suctions on North Elevation (kPa)*

The ASCE 7-98 Standard was used to compare the predicted pressure distribution from the wind tunnel tests; comparisons with wind tunnel predicted suctions are provided for enclosed and partially enclosed building conditions (Table 1). Both the wind tunnel test results and the ASCE 7-98 results assume an effective tributary area of glass of less than 1.9 m^2. Based on the zone definitions provided in the ASCE 7-98, approximately 90% of the total wall surface area of the building is considered a Central Zone, and 10% is considered a Corner Zone. In examining the wind tunnel test results, the recommended suction on approximately 90% of the total wall surface area is 1.4 kPa, on 9% of the building it is 1.9 kPa, and on 1% of the building the suction is 2.4 kPa.

Table 1 - *Comparison of ASCE and Wind Tunnel Derived Suctions for 50-year Return Period Cladding Design (kPa)*

ASCE 7-98 Analytical Method				Range of Wind Tunnel Results[3]
Enclosed Building[1]		Partially Enclosed Building[2]		
Central Zone (Zone 4)	Corner Zone (Zone 5)	Central Zone (Zone 4)	Corner Zone (Zone 5)	
1.4	2.5	1.9	3.0	1.4 to 2.4

[1] assumes uniformly distributed leakage
[2] assumes a dominant opening
[3] internal pressures determined using rational approach [*10*]

Note that using code analytical procedures, approximately 10% of the facade (the Corner Zone's) would have been designed for a suction of 3.0 kPa (assuming a partially enclosed building). The measured loads were 1.9 kPa for approximately 9% of the building facade and 2.4 kPa for 1% of the building facade. It is significant that the highest suction predicted from the wind tunnel study occurred in the middle of the curved north face of the building. Building code analytical methods would not have suggested this distribution of pressures.

The above comparison focuses on the suctions rather than the positive pressures since generally the highest cladding loads on a building are negative. The predicted positive pressures by the ASCE 7-98 and the wind tunnel tests agree quite well and are of similar magnitude to the central zone negative values.

It is important to note from this case study that the wind tunnel predicted lower wind loads on the facade than code analytical methods. The wind tunnel also predicted a wind pressure distribution that varied from that suggested by code.

Conclusions

Wind tunnel tests provide precise measurements of wind pressures on the cladding of buildings and structures by including the effects of irregular building geometry, terrain exposure and topography, and the influence of surrounding structures. The results of wind tunnel test methods can be used in place of code calculated loads. In the example presented in the paper there were notable differences between the magnitude and distribution of the wind tunnel predicted design wind suctions and the code derived suctions. The significance of this difference in the magnitude and distribution of the suctions is two-fold: 1) certain cladding elements on the curved face of the building may have been designed for only 80% of the actual specified wind load acting on them; and 2) cladding elements in the Corner Zones may have been designed for suctions up to 60%

greater than necessary (assuming a partially enclosed building). The wind tunnel method allows the cladding consultant to produce economical and risk consistent designs, which may result in significant savings for the client.

References

[1] ASCE 1998, "Minimum Design Loads for Buildings and Other Structures," American Society of Civil Engineers, ASCE 7-98 Standard.

[2] Cochran, L. S., and Peterka, J. A., "Building Shape and Cladding Loads: The influence of balconies and slab-edge storm shutters," *Wind Engineering into the 21st Century*, Balkema, Rotterdam, 1999, pp. 1117–1124.

[3] Blackmore, P. A., "Effects of Flow Channeling on Gable Wall Pressures," *Journal of Wind Engineering and Industrial Aerodynamics,* 38, 1991, pp. 311–323.

[4] Irwin, P. A., Cicci, M. D., and Lankin, J. B., "Variability of Cladding Pressures Caused by Adjacent Buildings," *Journal of Wind Engineering and Industrial Aerodynamics,* 77 & 78, 1998, pp. 147–156.

[5] Irwin, P. A., "The Design of Spires for Wind Simulation," *Journal of Wind Engineering and Industrial Aerodynamics,* 7, 1981, pp. 361–366.

[6] Irwin, P. A., and Kochanski, W. W., "Measurement of Structural Wind Loads Using the High Frequency Pressure Integration Method," *Proceedings of ASCE Structures Congress,* April, Boston, 1995, pp. 1631–1634.

[7] Irwin, P. A., "Pressure Model Techniques for Cladding Loads," *Journal of Wind Engineering and Industrial Aerodynamics,* 29 (1988), pp. 69–78.

[8] Davenport, A. G., "The Prediction of Risk Under Wind Loading," *Proceedings of the 21st Conference on Structural Safety and Reliability*, Munich, Germany, 19–21 September, 1977, pp. 511–538.

[9] Lepage, M. F., and Irwin, P. A., "A Technique for Combining Historical Wind Data with Wind Tunnel Tests to Predict Extreme Wind Loads," *Proceedings of the 5th US National Conference on Wind Engineering*, Lubbock, Texas, 6–8 November, 1985.

[10] Irwin, P. A., and Sifton, V. L., "Risk Considerations for Internal Pressures," *Journal of Wind Engineering and Industrial Aerodynamics,* 77 & 78, 1998, pp. 712–723.

[11] RWDI, "Final Report - Cladding Wind Load Study," Job. No. 99-450, Guelph, Ontario, Canada, 1999.

William H. McDonald[1], *and Michael D. Lewis R. A. M.S.C.E.*[2]

The Importance of Studying Exemplars when Designing Stone Facades

Reference: McDonald, W. H., and Lewis, M. D., **"The Importance of Studying Exemplars when Designing Stone Facades,"** *Performance of Exterior Building Walls, ASTM STP 1422*, P. G. Johnson, Ed., ASTM International, West Conshohocken, PA, 2003.

Abstract: The facades of contemporary buildings clad with dimension stone buildings all too often perform poorly, compared with the walls of much older buildings; despite they're being developed with computers and assembled using space-age parts. At the same time, structures using centuries-old methods often serve their occupants better than buildings many decades younger. Many facades built recently as architects' masterpieces now need major restoration simply to continue service. Their leaks, drafts, and dilapidated states shorten their useful lives.

Problems occur in new facades because their designers rarely understand the complex natural forces acting upon them. Earlier builders observed the performance of successful buildings, *exemplars*, and borrowed from them to suit their needs. In this way, the solid wall of antiquity evolved into a high architectural and structural art that has endured for centuries. The solid wall changed over the last hundred years to become cladding covering a separate structural skeleton. Although this concept offered aesthetic freedom and factory-expediency, a comparatively small portion of the last century's building inventory remains as healthy exemplars. Few of them have escaped major restoration efforts. And in the past fifty years, new material combinations spawned unprecedented problems. Leaks progressively attacked their concealed supports, inevitably damaging cladding and spiraling damage claims.

Contemporary builders repeat mistakes, partly because structural theory insists this procedure works, even as similarly constructed buildings fail. *Poorly*-performing exemplars are duplicated because their poor performance is ignored. Meanwhile, many existing older buildings, time-proven examples of sensible construction procedures and systems that perform well, continue to serve. In engineering's evolution from an *empirical* process to a *theoretical* one, stone façade construction has changed dramatically. Unfortunately, contemporary builders often ignore *exemplars* in a continuing but failed effort to fashion new images. To reduce expensive failures, successful builders study past performance. This paper examines the melding of empirical wisdom with new philosophies in producing useful buildings.

Keywords: architecture, construction, durability, exemplar, facade, stone, performance.

---

[1] Principal *(deceased 2001)*, The McDonald Group, Bedford, IN, USA.
[2] Consultant, Facade Forensics, Inc., 7739 Coldbrook Lane, Cincinnati, OH, 45255, USA.

Architecture's Departure from Exemplars Many Times Yields Failures

When designers begin a new project, they usually start with the building's appearance. When that new project is presented to clients, it is simple to show them their *objective* needs are fulfilled. They then debate *subjective* proposals for the exterior "look" of the new building. Heightening that project's purpose by making it handsome and unique will be a high priority. For this reason, the designer embellishes the building's exterior.

Often, in recent years, and continuing through the present time, stone has been used in decreased thicknesses. This new thinness, unthinkable in previous eras, made cladding lighter and *initially* less expensive. Designers believed and often abetted their client's expectations that thin walls would be economical and could successfully duplicate the appearance of strength and durability of the earlier stone buildings that inspired their looks. Borrowing images of older architecture without borrowing their construction methods left durability behind. Completely new styles and revivalist philosophy reborn in the early 20th century typically creates havoc on stone construction methods, as demonstrated when building owners routinely seek legal relief for building failures.

Unfortunately, most designs — indeed, from first image through nearly 90% of construction detailing — fail to address stresses that affect the wall stone panels system. Discovery of those stresses often occurs so late in the process that stability and durability are left to chance. Details such as an appropriate anchor concept to attach the stone to the wall is often delayed until shop drawings, or worse, is not engineered at all and instead are crudely fitted by the setter in the field.

This situation requires professionals of different disciplines to devise and coordinate separate systems to attach thin, *rigid* materials - glass, tile, stone, brick, metal – in the belief that more materials mixed in new ways creates better architecture. Separate support systems for mixed claddings become more and more complicated to fit into the thinner walls of these designs. Computer-driven *rational* analysis and drawing needed to configure the complicated systems has become the cornerstone (and crutch) of contemporary architectural design. On the other hand, some observers believe that a process aimed purposefully at accommodating the material to the image cannot be called rational. Endurance of *empirical* simple past practices that work must be a founding ingredient to any *rational* design. Failures ranging from falling stones, glass, and tile, to leaks and open joints tend to support the latter belief.

Exemplar Use Is Not New or Novel

This bleak scenario need not continue. A return to *empirical* wisdom from *less* enlightened times, when fewer problems developed, is the solution to reversing this dangerous trend. The design team should use widely available data regarding anchor systems and stone performance when conceiving its project's appearance. After conception, as an initial guide to develop a buildable scheme, they should use that information to study existing buildings. This is the *exemplar rule*.

Architecture Without Exemplars at the University of Cincinnati. The Vontz Center for Molecular Studies by Frank Gehry (1999) *top & bottom left* and Aronoff Center for Design by Peter Eisenmann (1996) *bottom right* create shapes and arrangements that have no precedent: their departure from proven form cannot benefit from the known durability of exemplars.

For hundreds of years buildings evolved and improved. The architect as the master builder studied existing structures, emulated their valuable characteristics, eliminated their obvious and dangerous defects, and improved their successful elements. The responsible contemporary design team can achieve the same results utilizing exemplars just as master builders did. Following the exemplar rule includes examining buildings of similar design, and service conditions, which utilize similar stones. It includes looking for clues in the buildings' performances in similar geographic and local exposures which will guide them in emulating, or avoiding, such architectural arrangements, support and anchor systems, stone types, sizes, finishes, or jointing, in their own building.

The exemplar rule is so simple, and so obvious, that design philosophies devoted to complexity all too often ignore or overlook it. When design teams try to make statements with their architecture, the utility, durability, and longevity of their creations are unavoidably secondary. Spending so much time inventing *image* seems not to leave enough time to develop construction methods, properly adapt known methods or even research precedents for the compilations they're conceiving. Ignorant of real-world-real-time performance, the laboratory process has established that theoretical engineering analysis and small specimen testing predict large results. The people following the process refuse to consider clear, more accurate, and easily available evidence just across the street or across town.

Finding Exemplars

Exemplars may not be easily available for seldom-used stones or novel support systems aimed at fitting the shapes and arrangements of stone on the new building. In those cases, if the designer chooses to use an unproven stone instead of one of the countless stones that have exemplars, he may have no choice but to depend on expensive and time-consuming weathering and full-size structural testing. Still, in a remarkable number of cases where exemplars *are available*, decisions on building skin durability are based on small-scale testing without any regard to past performance.

But the vast majority of stones available from U.S. production, and many stones from foreign quarries, have been in constant use for decades in all areas of the U.S. For the designer considering one or more of these, there exist buildings with sufficient satisfactory service history to illustrate the durability of the stone and its support. Often, such exemplars will prove to have anchor and support schemes similar in type to those under consideration.

If no exemplar can be found, the wise designer must wonder if using a stone or system without precedent is worth the risk of not knowing time-proven performance. Visionary designers more interested in cutting-edge fashion often put themselves in this quandary. If it fails to provide shelter and high function, then it fails as architecture and is only art. Infatuated with fashion and signature identities, some architects fail to even know when their construction competency is lacking. The largest percentage of failures occur in this type work, compared to more traditional designs. This opens the question of what determines excellence in architecture: critical acclaim or function without failure. Most building owners want minimum failures.

Limestone Exemplars at Indiana University. Maxwell Hall (1895) *top* where century-old tooled texture on watertable is still crisp. Student Building Clock Tower (1902) *bottom left* and Music Building (1940) *bottom right* where testing showed the strength of exposed, weathered stone equaled or exceeded the strength of unexposed surfaces of the same panel.

Exemplars...Good and Bad

Universities, state and local governments, religious and other organizations, which build multiple buildings within a campus environment, can be particularly good places to find exemplars. Here one can often find examples of good construction and material use, as well as blunders to be avoided. Both *well*-performing and *poorly*-performing works are *good exemplars*. This is particularly true of those campuses where a certain material is used over and over again. That material can become the thread of identity, which unifies a campus' style of architecture across different types of construction. Built over different periods, usually attempting to be durable, the condition of their works could show a broad spectrum of methods, even with similar appearances. Ohio State University is one example of similar style buildings constructed by methods evolved over more than a century. Its mid-city campus, in state capitol Columbus, is of red brick with limestone trim. The school's decision-makers continue the solid and dependable buildings such construction yielded, both in appearance and in performance. Yet the durability of early load-bearing and later veneer structures is very different, particularly those only a decade old. University of Chicago structures span styles from Gothic, through Brutalism, to Post-Modern with its limestone buildings. University of Cincinnati represents the same era with a multitude of styles and nearly every construction material and method. While finding examples of different construction is easy, understanding this engineering diversity requires some research. Interpreting good and poor performance requires comprehensive knowledge of materials and methods over history to know how hidden construction beneath the surface was generally built.

One example of a poor-performing structure serving as an excellent exemplar is a landmark building in Columbus, the Ohio Departments Building. Finished in 1932, clad in white Georgia Marble, the building suffered sixty-five years of neglect and poorly planned repairs. Most observers believed that the deteriorating marble was at fault, in spite of its many years' successful use in service conditions far more severe than found in Midwest America. Casual investigation proved insufficient and failing support ruined the marble. It was a wonder the stone withstood the symptoms of that era's transition from load-bearing to lighter veneer walls on tall buildings. Successful long-term rehabilitation required recladding to expose, redesign and construct a modern support for the marble cladding to correct the original construction's problems.

Another example of exemplar usage developed as a result of an entirely different intention. Beginning in the 1960s, ignoring its more than one hundred years of successful duty on structures nearly everywhere in the nation, some thought Indiana Limestone and all calcareous stones deteriorated in acid precipitation. As the trade association for the material, Indiana Limestone Institute (ILI) assisted by the Indiana Geological Survey (IGS) researched the extent of damage in southern New England, the area of the U.S. most affected by acid precipitation. The extensive study of limestone buildings published in 1986 revealed that, contrary to first thought, limestone and marble specimens from the subject buildings suffered little if any damage from acid precipitation. After visually inspecting buildings for distress, and obtaining specimens of weathered and non-weathered stones from the same buildings, ILI and IGS tested and compared the stones' resistance to crushing and rupture (ASTM Test Methods C170, and C99 respectively). Ironically, tests showed certain calcareous stone

Projects Using Precedents. Ohio Departments Building (1932 original, renovated 1999) *top* reclad with same type marble but on new type support after the original support failed. State Teachers (1985, 2000) *bottom left* supports addition's limestone on trusses. Cleveland Clinic Cancer Center (1999 by Cesar Pelli) *bottom right* puts stone on unitized curtainwall.

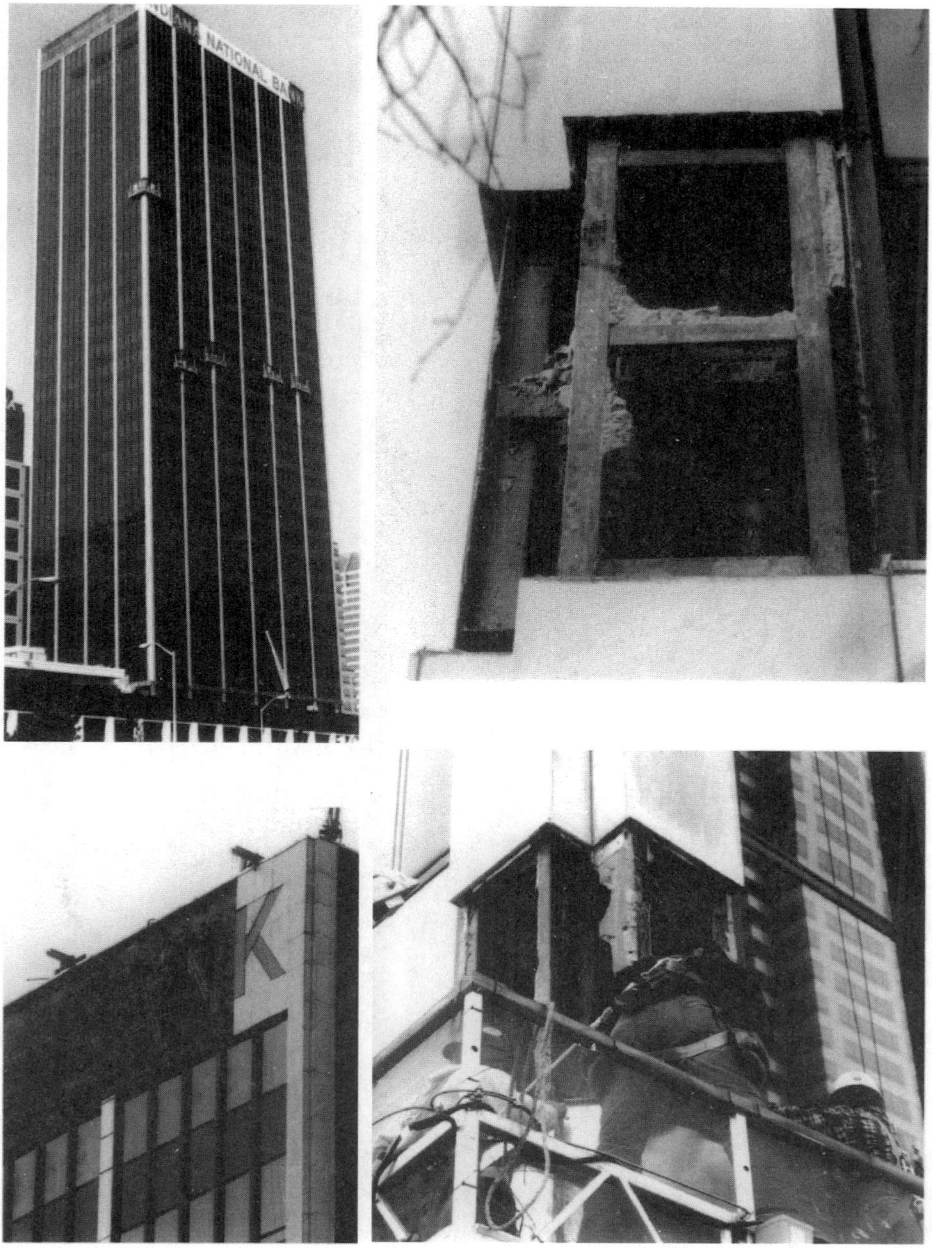

Unused Exemplar. Indiana National Bank (1960) originally clad in Carrara marble showed bowing and distress soon after its 1969 completion, while Standard Oil (later Amoco) was still on the boards. Remedial restraints placed in the early 1970s did not stop panel movement. All panels were eventually removed when INB was finally reclad in 1992.

Chicago's Amoco Is Stripped of its Original Carrara Marble and Reclad in Granite. Only twelve years after its 1973 opening, periodic inspections found panels starting to bow outwards. By 1987, some of the 1¼ in. to 1½ in. thick 50 in. by 45 in. marble panels had bowed 1½ in.. More than 44,000 panels were replaced while occupants worked inside.

strengths actually increased when weather-exposed for many years when compared to unexposed stones of similar age.

Results were duplicated on three Indiana University buildings in Bloomington. Specimens were prepared to compare weathered surfaces in bending resistance to unweathered sections *from the same stones*. The weathered specimens were equal in strength or stronger than unweathered ones in bending and crushing. Although this has not been scientifically explained, some geologists believe strengthening could result from migration of calcium toward weathering surfaces from below the surface due to repeated wetting and drying from natural exposure, referred to as "case hardening." The findings did not surprise Indiana quarrymen, who for decades have maintained that their stone hardens on exposure; citing more difficult cutting, shaping and carving of well-weathered blocks.

Investigating limestone exemplars proved valuable to owners of buildings constructed using certain limestone and marbles, and to architects planning new projects contemplating those materials. It also proved that many calcareous stones, sometimes viewed as inferior to stones with higher test values, were dependable. Examining entire exterior wall systems of existing buildings like those in Cincinnati, Columbus and Chicago, for prospective projects in those environments, would be equally valuable in determining dependable building methods.

The Amoco Building, Where a Search for Exemplars Could Have Helped

Near Chicago's lakeshore, the Amoco Building, completed in 1973, stood for years as a stark white beacon, one of Chicago's posh addresses. Then, early in the 1980s, it was discovered that many of its Carrara marble panels were bowing distressingly outward. The 50-by-45 inch panels, not excessively large, but their 1¼ to 1½ inch (3 to <4cm) thickness was, and still is, distinctly *non*-conservative for marble work. While repair was compared to replacement, stainless steel bands were snugged around the most egregiously-bowed panels.

Meanwhile, after months of study and tests, teams of consultants, engineers, architects, lawyers, and mechanics eventually recommended that the most prudent and economical permanent correction would be to replace every marble panel. Panels that are 50% thicker, this time of North Carolina white granite, attached by continuous extruded stainless steel anchors, comprise the 670,000sf reclad facade. An estimated $75 million (approximately $112/sf) and three years later, recladding was accomplished, and the building was again deemed safe to walk or drive around. The recladding construction contract cost nearly 60% of the original building's construction, excluding payments to lawyers, engineers and consultants.

Experts disagree about what physical processes cause some marble panels to deform. Other Carrara-clad projects in Indianapolis, Denver and Helsinki met the same fate as Amoco. The consensus opinion believes bowing is residual dilation resulting from anisotropic thermal expansion of the extremely fine <0.3mm grains. Simply, inside and outside faces of the panel don't expand equally, then don't shrink completely back when the temperature returns. From this point, scientists debate whether vapor drive or stress from loads increase bowing. A good vapor barrier and individually supported panels make this debate moot.

On the thirty-six-story Indiana National Bank, Carrara marble column covers began to show bowing and deformation soon after its completion in 1969, before Amoco began. The first fix attempt in the early 1970s face-drilled to re-anchor the tower panels to the structure, however the marble panels continued to deform even after being restrained. Conden and Lamson, Indianapolis architects ultimately designed and oversaw replacement of the marble entirely with a metal system. However, before either INB or Amoco, expert geologists had observed and documented the distortion phenomenon in Carrara-type marbles as early as 1919.

The Value of Hands-On

An experienced quarrier or stone fabricator can often predict his product's strength characteristics almost close enough to do preliminary engineering simply by handling spalls from a quarry block. The experienced stone engineer can reasonably diagnose internal causes of distress from symptoms seen at the surface by hands-on examination of signs of movement, cracks, spalls, restrained areas, joints, and openings. The hands-on approach teaches the invaluable empirical knowledge that would prevent problems, invisible on paper and in small laboratory specimens, from repeating. It is not difficult to acquire such knowledge and skill, but first, one has to be convinced the complex, concurrent atmospheric influences of natural exposure over *real* time is a better predictor of durability than *assumed* accelerated conditions in a laboratory. Understand the true conditions the building must provide service, and that they vary on the same building, even the same façade, depending upon *many* elements.

By intelligent and diligent inquiry, stone users and specifiers can learn about the stones they may be considering by consulting this knowledge and by practicing hands-on. When rubbed, does weathered stone from the same source seem to powder away? Are building arrises still sharp? Is ornament and lettering crisp? Major surface loss, or blurred lettering, or variable finish, over relatively few years, should be reason for serious and intensive examination. Is distress concentrated at certain areas such as corners or where anchors might be while other areas are relatively intact? How has building structure and stone support contributed to the symptoms at the surface? Does the stone cladding material, its support, the building frame, or the three not being compatible cause the problems?

This sort of inquiry is the beginning of the *exemplar rule*. The user or specifier must see not only how the stone behaves, but how it interacts with its support system, its joint closures, and its anchors. Using the same discerning eye one has while compiling a punchlist, most interested observers can collect the same information on an exemplar before design begins as one does to finish a project, then capitalize on the findings. To *empirically* analyze stone on exemplar buildings:

- Examine the chosen stone's structural properties as illustrated by panel sizes, thicknesses and support points;
- Assess realistic production limitations such as piece sizes; stone quantities (Can the quarry produce the project's volume of stone? Can the fabricator meet delivery schedule?);
- Determine range of natural characteristics such as color, stone texture, occurrence of natural, though perhaps undesirable coloration or graining conditions, stability and

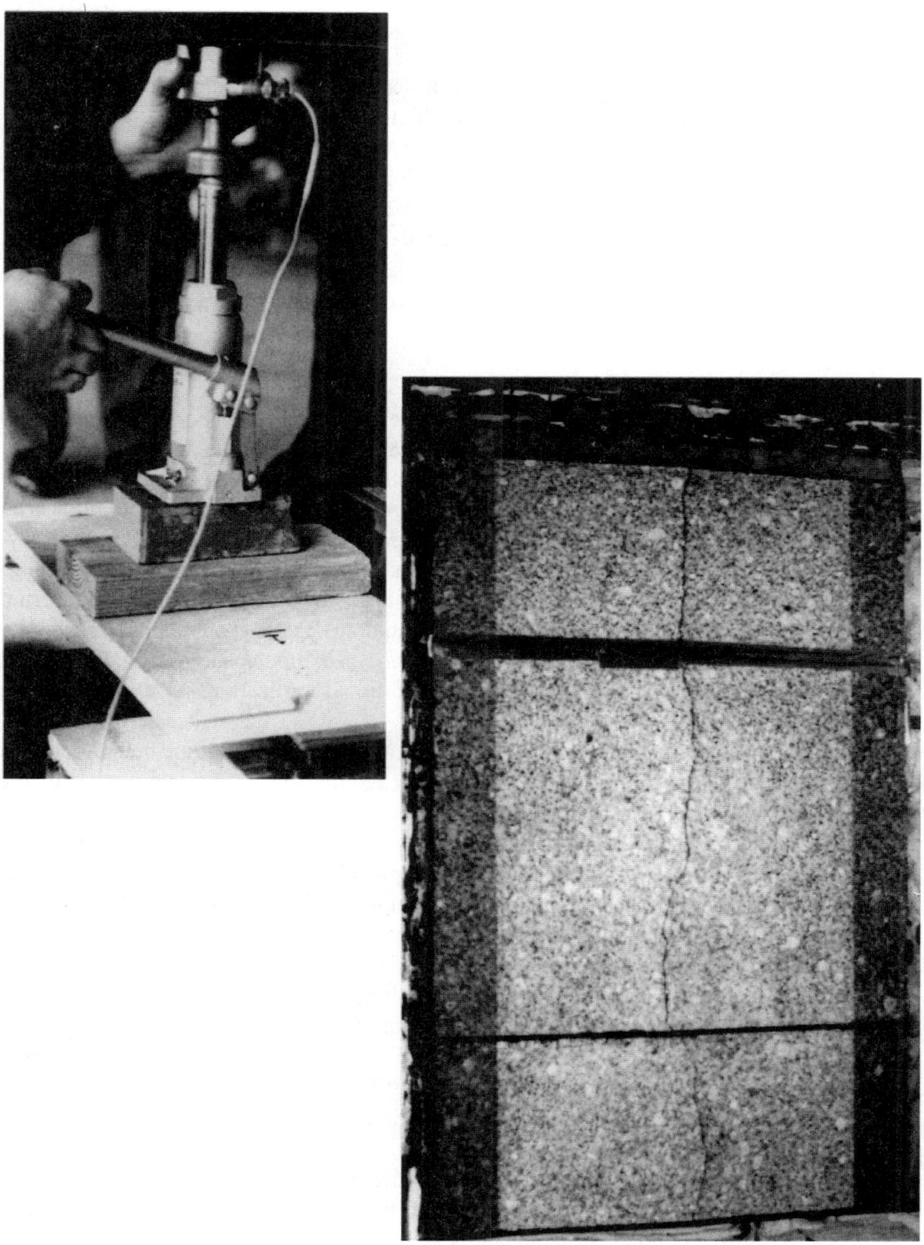

Simple Tests Predict Structural Capacities, But Not Durability. One version of ASTM's C 1254 *top* isolates the capacity of the stone/anchor engagement. ASTM's C1201 *bottom* proves assembly integrity using pressure differential to measure panel capacity in plate bending on its anchors connected to its backup that deflects as it will in the project.

evenness of finish, and structural capability of desired variance of materials;

- Inspect for symptoms of problems inside the wall and their possible explanations: stains, cracks, spalls, bulges, open joints, shifted stones, varying sight lines along string courses, bull noses, copings, sill runs, building corners and the like, plane changes, window and other opening intersections with stone.
- Inspect for problem fit conditions, such as lippage, warping planes, uneven corners, tapering joints, and other conditions, which may indicate fabrication deficiencies, setting problems, inadequate anchoring, or poorly matching structure or backup.

Use Your Calculator

The designer can also gain a good knowledge of probable anchor performance by comparing his own or his engineer's anchor scheme with the examples of anchor characteristics described in ASTM's C1242, *Standard Guide for Selection, Design and Installation of Exterior Dimension Stone Anchors and Anchoring Systems*. The Guide is an overview of techniques, and a review of procedures orienting the designer to industry-consensus sound construction techniques. ASTM's Manual 21, *Modern Stone Cladding: Design and Installation of Exterior Stone Systems,* leads the designer through engineering methods and how to fit cladding and its support to buildings. Design handbooks published by stone trade associations, such as the *Indiana Limestone Handbook* and Marble Institute of America's *Design Manual,* illustrate typical and traditional applications. Contained in those sources are procedures to:

- Assure stone producers' reports show that the properties of the stone being produced for the project exceed the design minimums required by engineering needs of the project, not simply generic specifications;
- Compare published strength data with stresses factored for safety. The factor must consider the overall system, exposure and application, not simply and only the stone material. If comparison fails, test stone samples before production. If the project warrants, continue testing through production to check consistency;
- Analyze stone where it engages its anchor device and analyze the device itself to confirm interaction capacities (lateral only? lateral plus gravity?) under conditions that will occur on the project over the expected life of the project;
- In the case of large panels, or those supported in a way that subjects the panel to two-way bending, test assemblies using ASTM Test Method C1201 to confirm compliance with stresses and to assure that the system's deflections do not induce added stresses unaccounted in the structural analysis;
- Analyze supporting backup for accommodation to movement and deflection where anchors connect, and where backup connects to building. Load path must be checked back to primary building structure and all cumulative effects accommodated.

Matching a proven stone, in sizes appropriate to resist the expected stresses and exposure, with an anchoring system known to be capable of supporting the skin, is the essence of the *exemplar rule* in stone architecture. More stone users should follow it.

SECTION II

Bruce S. Kaskel[1] *and Thomas R. Wegener*[2]

"Building a Better Wall System"; The Application of ASTM E 2099 "Standard Practice for the Specification and Evaluation of Pre-Construction Laboratory Mockups of Exterior Wall Systems"

Reference: Kaskel, B. S., Wegener, T. R., **"Building a Better Wall System"; The Application of ASTM E 2099 "Standard Practice for the Specification and Evaluation of Pre-Construction Laboratory Mockups of Exterior Wall Systems"** *Performance of Exterior Building Walls, ASTM STP 1422,* P.G. Johnson, Ed., ASTM International, West Conshohocken, PA, 2003.

Abstract: ASTM subcommittee E06.55 has recently developed E 2099 "Standard Practice for the Specification and Evaluation of Pre-Construction Laboratory Mockups of Exterior Wall Systems." This practice has been developed to standardize the process for the design, construction, and laboratory testing of full-scale pre-construction mockups. These mockups are typically considered standard practice for buildings where the exterior wall system is a custom application and for buildings where a high confidence level is desired in the performance of the exterior wall system. To date, laboratory mockups have been tested using ASTM E 283, E 330, E 331, E 547, and E 1233, as well as tests developed by other organizations. E 2099 brings these various individual tests into one coherent test program. E 2099 also provides specific recommendations for the exchange of information between the exterior wall designer, the exterior wall contractor and the test laboratory. The intended goal of such an exchange of information is to provide a mechanism for valuable lessons learned in the process of the design, construction and testing of the mockup to be carried through to the product installed on the building.

Keywords: laboratory mockups, pre-construction evaluation, exterior wall systems, testing, curtain wall

Introduction

Exterior walls of today's buildings are more complicated and must meet more demanding requirements than ever before. Facades must not only be watertight and wind resistant, but must accommodate thermal expansion and contraction, meet stringent energy performance requirements, prove durable over a long life span, and provide safe

---

1 Senior Consultant, Wiss, Janney, Elstner Associates, Inc., 120 N. LaSalle Street, Suite 2000, Chicago, IL 60602.

2 Senior Architect, Wiss, Janney, Elstner Associates, Inc., 120 N. LaSalle Street, Suite 2000, Chicago, IL 60602.

conditions even in seismic events. At the same time, architectural facade designs have become more complex, using more materials and novel systems than ever before. Because of all these condition, facades require careful design. But even the best designs still demand some method to verify facade performance. Testing full-scale laboratory mockups of the exterior wall system prior to construction can contribute to performance verification. Such testing is today normally considered for buildings where the curtain wall is either a custom application, a high-rise building, or a building where a high confidence level is desired in the performance of the curtain wall. [1].

Johnson identified mockup testing as an important part of quality assurance in the building process. As he noted:

> "Many exterior wall systems will require mockup or performance testing to confirm the wall design will conform to the particular appearance and performance criteria of the project. This process is especially important to those wall systems with appearance, design, or performance criteria which are unusual, or which have not been previously constructed. After exterior walls are in place, repairs can be expensive and difficult." [2]

Johnson advocated a "Quality Assurance Program" (QAP) that carried through the entire design and construction process, including the mockup testing. Johnson noted the critical nature of scheduling the mockup into the design/construction process:

> "Keeping the mockup and performance testing on the proper schedule track can be a critical link in avoiding water infiltration problems in the completed wall. As in many other areas of construction, if the mockup or performance testing does not keep pace with the overall project schedule, there is likely to be pressure to short cut the process. If this pressure is successful, a valuable tool in discovering potential sources of water infiltration is lost". [2]

History of Mockup Test Standards

The current state-of-the-art in exterior wall mockup testing had its origin in the post-World War II development of metal and glass curtain wall systems. The need to develop standard tests applicable to curtain walls was recognized by Hunt in 1958:

> "The testing of curtain wall assemblies has not been very well organized. For the most part, it has been based on tests originally developed for other products, such as windows or experimental panels for low-cost houses. Adding to the confusion is the lack of accepted criteria and the almost complete dearth of assembly standards". [3]

Hunt further stated that:

> "Tests and standards do exist, however, for many components. These are valuable as far as they go and should be used. Yet even for many of the components for which valid tests exist, no criteria have been generally

accepted for applying them to curtain walls. This situation has left many conscientious curtain wall suppliers in the peculiar position of having to determine whether their products are acceptable without knowing what constitutes an acceptable product". [3]

At the same time that Hunt was lamenting the lack of curtain wall tests, standard test methods were being developed abroad. Sakhnovsky [4] reports on the 1950 research of the Norwegian Building Research Institute that developed a method of testing windows: "Using an elaborate apparatus, including a chamber covered by zinc with soldiered joints and a water application system of air feeding streams of water into jets traveling in a prearranged cycle." This led, in the early 1960s in the US, to static pressure window testing using a "double chamber method," as recommended by the National Association of Architectural Metal Manufacturers (NAAMM). According to Sakhnovsky, currently in the United States, double chambers are used "only for thermal testing, where a different environment is necessary on each side of the wall for the thermal-testing program." There are several standard tests intended to determine thermal properties of exterior wall components such as "U" value and condensation resistance, which require double thermal chambers, known colloquially as "hot boxes". These tests, which have specific size requirements smaller than typical laboratory specimen sizes for the test procedures discussed below. These thermal tests are commonly performed once for each standard manufacturer's system and not on pre-construction mockups.

ASTM Standards

During the 1960s, ASTM introduced three standards that serve to this day as the central tests procedures for exterior wall system mockups (italics added for emphasis):

- ASTM E 283 Standard Test Method for Determining the Rate of *Air Leakage* Through Exterior Windows, Curtain Walls, and Doors Under Specified Pressure Differences Across the Specimen
- ASTM E 330 Standard Test Method for *Structural Performance* of Exterior Windows, Curtain Walls, and Doors by Uniform Static Air Pressure Difference
- ASTM E 331 Standard Test Method for *Water Penetration* of Exterior Windows, Curtain Walls, and Doors by Uniform Static Air Pressure Difference

All three tests were conceived so that they could be performed on the same mockup specimen. All three tests rely upon a "single chamber"; i.e. the sealed test chamber constructed solely on one side (usually the interior side) of the mockup test specimen, in order to create an air pressure difference between the "interior" and the "exterior" ambient air pressure. The static air pressure difference is intended to represent the effects of wind load on exterior building surface elements. Section 5 of E 330 provides a complete discussion of the significance and use of the static air pressure difference concept. The user of standard E 2099 is recommended to read all three standards for an understanding of these test methods.

Since the 1960s, ASTM has introduced two test methods that are applicable to exterior wall mockups which have been included in standard E 2099:

- ASTM E547 Standard Test Method for *Water Penetration* of Exterior Windows, Curtain Walls, and Doors by *Cyclic* Static Air Pressure Differential (Originally published in 1975)
- ASTM E1233 Standard Test Method for *Structural Performance* of Exterior Curtain Walls, and Doors by *Cyclic* Static Air Pressure Differential (Originally published in 1988)

These tests differ from the original three largely in their use of pressure cycles in lieu of a single static pressure difference. This modification is intended to allow a more accurate representation of complex wind loading. E547 commonly is conducted with low cycles (three, 5 minute cycles is common), while E1233 has non-mandatory provisions for high-cycle tests intended for extreme winds and hurricanes.

Recently, ASTM has introduced the following two standards applicable to exterior wall mockups. These two standards have been referenced by the International Building Code 2000, as the standards for wall systems in windborne debris regions; generally the costal regions of the Atlantic Ocean, Gulf of Mexico, Alaska, Hawaii, and U.S. Pacific territories:

- ASTM E1886 Standard Test Method for Performance of Exterior Windows, Curtain Walls, Doors, and Storm Shutters Impacted by Missile(s) and Exposed to Cyclic Pressure Differentials (Originally published in 1997)
- ASTM E1996 Standard Specification for Performance of Exterior Windows, Curtain Walls, Doors and Storm Shutters Impacted by Windborne Debris in Huricanes (Originally published in 1999)

Other Standards

The American Architectural Manufacturer's Association (AAMA) also promotes standard test methods that are commonly performed on mockups. These are:

- AAMA 501.1, Standard Test Method for Exterior Windows, Curtain Walls and Doors for *Water Penetration Using Dynamic Pressure*
- AAMA 501.4, Recommended Static Test Method for Evaluating Curtain Wall and Storefront Systems Subjected to *Seismic and Wind Induced Interstory Drifts*
- AAMA 501.5, Test Method for *Thermal Cycling* of Exterior Walls

The first of these three, the "dynamic pressure" test, simulates wind and rain conditions using a large propeller aircraft engine wind generator and water spray in lieu of the water plus a static pressure difference. As noted in the test method [5]. "Because of the turbulence, sources of water penetration may be found which would not show up in the uniform static air pressure test. The dynamic method is considered to more closely represent the impact caused by unpredictable and suddenly shifting wind gusts and wind-blown water." Sakhnovsky [4] notes that, "Despite some early opposition, 35 years of testing have demonstrated some significant advantages to the addition of the dynamic test method".

Furthermore, non-standardized tests such as scaffold stabilizer pull tests and tests to stimulate the effect of deterioration of the seals [1] are occasionally also performed on the mockup.

New Standard Practice ASTM E 2099

ASTM subcommittee E06.55 on Exterior Wall Systems, initiated in 1996 a task group to develop a new standard practice for the testing of exterior wall systems. The original statement of purpose for this task group was:

> "It has become commonplace that the curtain walls of large buildings are tested for air infiltration, water penetration, and structural reliability prior to final design, fabrication, and installation at the building. Properly designed, executed, and documented, mockup tests provide an excellent means to help establish the reliability of the initial installation, and to provide a 'dry-run' to detect potential process problems such as long lead-time for materials, coordination and sequencing of trades, and fit-up or tolerance issues. Improperly designed, executed, or documented mockup tests can have a detrimental effect on the performance of the final wall construction by providing misleading indications of probable performance, and providing a false sense of security to the parties to the construction".
> A test protocol standard might address the following issues:
>
> - Specimen Selection – e.g., testing representative and critical elements of the wall system.
> - Documentation – e.g., record-keeping to help assure that the lessons are not lost, and that modifications made during the mockup phase are reflected on the final installation.
> - Selection of Appropriate Test Methods and Loads (this subject may require a separate standard).
> - Failure/Remediation Process – i.e., what to do when the mockup fails.
> - Mockup testing is a useful tool that has too often been misused. A collection of guidelines based on past experience can help avoid many of the pitfalls that have plagued past projects[3].

From this statement of purpose, the subcommittee developed ASTM E 2099, "Standard Practice for the Specification and Evaluation of Pre-Construction Laboratory Mockups of Exterior Wall Systems." This standard was adopted by ASTM in 2000.

The standard recognizes that a mockup test requires the coordinated work of three principal parties; the specifier (or designer), the builder and the test agency. Each

---

[3] Letter from subcommittee chair to task group chair, October 22, 1996.

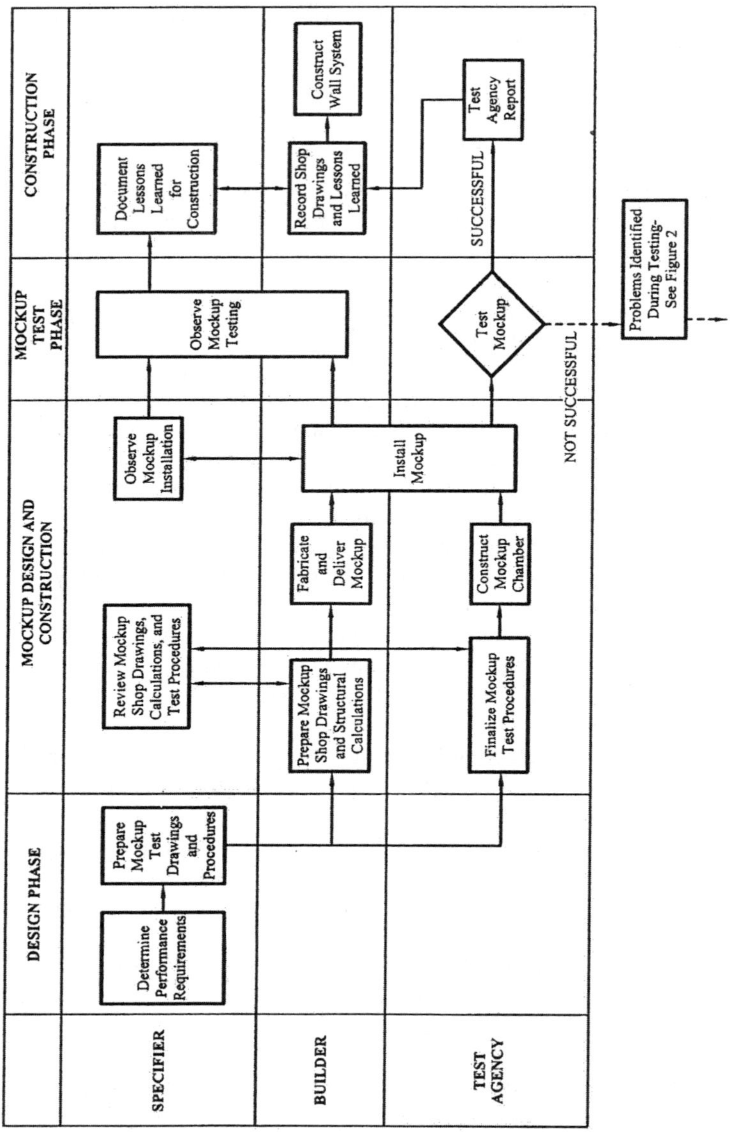

Figure 1 --*Flowchart of exterior wall system mockup process (adapted from Ref. [1])*

party has specific roles that should be carried out and information to exchange between parties.

Figure 1 shows a flowchart of the roles and interactions of the parties. E 2099 recognizes that contractual responsibilities may be assigned differently by project contract documents. For instance, in design/build projects the builder would also be the designer. In general, however, the three parties; specifier, builder and test agency, perform the following tasks:

1. *Design the exterior wall system*, to be described in plan, elevation, section and details. The design needs to be developed to a buildable level and should include performance requirements, usually in specification format. This is usually achieved at the 35% or 50% construction document phase. Section 5 of the standard describes the specifier's tasks in the design of the mockup program.
2. *Construct a full-scale portion of the exterior wall system,* which represents a portion of the typical or repetitive elements of the exterior wall design. This mockup can be constructed either at one of several testing laboratories within the US, Canada or abroad. Section 6 of the standard describes the builder's tasks in the construction of the mockup.
3. *Implement a series of mockup tests*, commonly performed by testing personnel with the necessary specialized test equipment. Section 7 describes the test agency's tasks in the testing of the mockup.

The following discusses the new E 2099 standard by these major sections:

ASTM E 2099 - Section 5. Design

Section 5 is the lengthiest portion of the standard. The architect or the professional who specifies the performance of the exterior wall system should consider the guidelines presented in this section. The section provides guidance on the following aspects of the mockup design:

- The requirements of the mockup size, materials and configuration
- The performance requirements of the mockup testing

Both of these are important to specify in order for bidders to understand the curtain wall requirements and to appropriately plan the mockup. When the mockup test performance requirements are not adequately specified, disputes may arise regarding the design intent. These disputes can often have significant cost implications if the builder did not anticipate the same level of performance required by the specifier.

The selection of the mockup size and configuration is discussed in the preceding ASTM standards, especially in the structural performance test E 330. The new E 2099 standard provides additional guidelines for the mockup size and configuration that reflects the types of exterior wall systems commonly built today. It is a basic understanding that for structural tests the mockup structural system should closely represent the system used on the façade especially in regards to strength and stiffness characteristics. E 330, although applicable to metal, glass, masonry and stone components, illustrates a curtain wall built with evenly spaced vertical mullions, horizontal rails, and regular-sized vision and spandrel units. As shown in E 330, the mockup would represent a typical two-story vertical mullion system attached at two floor slabs. This configuration would not work for many of the structural systems currently used in exterior wall systems, such as spandrel strong-back truss frames or panelized wall systems. Nor does it represent the mockup requirements for a façade constructed of several different wall materials. Section 5 of E 2099 provides guidance on how to select an appropriate mockup configuration given the myriad of exterior wall systems used today.

The subject of a mockup test order has also been covered in guides over the years [8], but has not been standardized through ASTM. The ASTM test methods such as E 330, E 331 and E 283 each had their own test procedure and pass/fail criteria. Although the specifier would normally stipulate the pass/fail requirement for each test, with multiple tests it is often easy to overlook critical information. In such cases, it usually falls on the test agency to supply missing information. It is not uncommon for the test agency to provide an order for the tests or even to recommend test load levels. E 2099 provides a checklist of necessary specification information to be provided by the specifier. Table 1 provides a list of specification information required in the tests references in the E 2099 standard.

E 2099 also provides the following default test order in cases where the specifier has not stipulated the order:

1. For mockups that have operating windows and doors, the first step is to open and close them at least 5 cycles. This is done to verify their performance and to free-up gaskets and weatherstrips. Any performance problems need to be resolved prior to testing.
2. Test 1 – E 330 (Structural Performance) to -50% of design positive test load. This step is intended to "preset" the mockup allowing for movements that can flex joints and stress gaskets and seals as would be found in actual conditions.
3. Test 2 – E 283 (Air Infiltration) at specified air pressure difference. Traditionally in the United States this test is performed at either 0.075kPa (1.56 psf) or 0.300 kPa (6.24 psf). These pressures were selected historically to represent the static pressure difference caused by a steady 40.2 kph or 80.4 kph (25 mph or 50 mph) wind respectively. This is typically performed prior to water testing because water trapped with in the wall tends to reduce air leakage [7]. A second air infiltration test is recommended subsequent to testing for interstory drift or thermal cycling to check their affect on air infiltration resistance of the wall.
4. Test 3 - E 331 (Water Penetration) at specified air pressure difference. Traditionally in the United States the specified air pressure difference is selected as 15 to 20% of the design positive wind pressure. Testing history has shown that pressures in this range create a severe test environment for the mockup which will not likely be exceeded during the building's lifespan. Most standards limit the lower end of test pressures to 0.137 kPa (2.86 psf). It is not uncommon for a pressure difference of up to 0.718 kPa (15.0 psf) to be specified.
5. Test 4 through 8 are optional tests and should be specified if desired.

Table 1 – *Information Provided by the Specifier*

TEST	INFORMATION TO BE SPECIFIED [1]	DEFAULT IN TEST METHOD [1,2]
ASTM E283 –91 (Reapproved 99) Air Leakage	1. Inclusion of window/door/other component in specimen (§ 8.2) 2. Specimen size (§ 10.1.1) 3. Pressure difference (§ 10.1.2) 4. (10.1.3) Air flow direction (exfiltration/infiltration) 5. (§ 10.1.4) Air leakage rate	3. 0.075 kPa (1.56 psf) 4. Infiltration 5. *Refer to AAMA 501 "Laboratory Test Specification" item 1, for recommendations.*
ASTM E330 –97 Structural – Uniform Pressure	1. Testing of anchorage system (§8.1.2) 2. Inclusion of window/door/other component in specimen (§ 8.2) 3. Specimen size (§ 8.2.1) 4. Testing of roof coping (§8.2.1-Note 3) 5. Additional information contained in report (§ 12.1.17) ***Select* Procedure A *or* Procedure B** **Procedure A** 1. Positive and negative loads (§10.1.1) 2. Duration of maximum load (§10.1.2) 3. Number and location of deflection measurements required. (§10.1.3) **Procedure B** 1. Number of incremental loads and load at each increment (§ 6.2.4.1 and 10.2.1) 2. (10.2.2) Duration at each load increment and at maximum load 3. Number and location of deflection measurements required (§ 6.2.4.2 and 10.2.3)	1. No testing 4. No testing 5. None 2. Not required but typically 1.5 x design load (§5.4). Minimum 10 seconds (§ 5.4 and 11.2.4) 3. None (§ 10.1.3) 1. Four approximately equal increments to maximum load (§ 11.3.3) 2. Minimum 10 seconds and 1.5 x design load (§ 5.4 and 11.3.2 and 11.3.3) 3. Three locations – maximum and @ end of principal member (§ 6.2.4.2)
ASTM E331 –00 Water – Uniform Pressure	1. Specimen size (§ 8.2.1) 2. Pressure difference (§ 10.1) 3. Failure criteria (§ 10.2)	2. 0.137 kPa (2.86 psf). *Refer to AAMA 501 "Laboratory Test Specification" item 2, for recommendations* 3. "Penetration of water beyond the vertical plane intersecting the innermost projection of the test specimen, not including interior trim and hardware, …" (3.2.3)

Table 1 (cont.) – *Information Provided by the Specifier*

TEST	INFORMATION TO BE SPECIFIED [1]	DEFAULT IN TEST METHOD [1, 2]
ASTM E547 –00 Water – Cyclic Pressure	1. Specimen size (§ 8.2.1) 2. Pressure difference (§ 10.1) 3. Failure criteria (§ 10.2) 4. Time duration of test (§10.3) 5. Temperature conditions (§10.4)	2. 137 Pa (2.86 psf) 3. See sections 3.2.3 and 10.2 for definitions. 4. Minimum Required – Two cycles with 5 minute minimum/cycle; 15 minimum 5. Total test length (§ 10.3.1 and 10.3.2) 6. Ambient
ASTM E1233 -00 Structural-Cyclic Pressure	1. Testing of anchorage system (§ 8.1.2) 2. Inclusion of window/door/other component in specimen (§ 8.2) 3. Specimen size (§ 8.2.1) 4. Testing of roof coping (§ 8.2.1-Note 4) 5. Additional information contained in report (§ 12.1.18) ***Select* Procedure A *or* Procedure B** **Procedure A** 1. Life cycle load (§ 10.1.1) **Procedure B** 1. Wind event load procedure (§10.1.1) **Procedure A & B** 2. Positive and negative cyclic load (§ 10.1.1.1) 3. Number and duration of cycles (§ 10.1.1.2) 4. Point in load sequence for deflection measurement and observation (§ 10.1.1.3) 5. Maximum positive and negative load (§10.1.1.4) 6. Duration of maximum loads (§ 10.1.1.5) 7. Number and location of deflection measurements required, if any (§ 10.1.1.6)	1. No testing 4. No testing 5. None See E1233 Appendix X1 for recommended guidance 5. 1.5 x design load (§ 5.9) 6. 10 sec. minimum (§ 5.9 and 11.5.2)

Table 1 (cont.) – *Information Provided by the Specifier*

TEST	INFORMATION TO BE SPECIFIED [1]	DEFAULT IN TEST METHOD [1, 2]
AAMA 501.1 -94 Dynamic Pressure	1. Equivalent static air pressure (§ AAMA 501 – "Laboratory Test Specimen," item 4)	1. *Refer to AAMA 501 "Laboratory Test Specification" item 2, for recommendations.*
AAMA 501.4 –00 Interstory Drift	1. Specimen size (§ 6.1)	1. Minimum width– two typical units including one typical vertical joint or framing member on both. Minimum height-full building story, or unit, or full mullion length, and horizontal expansion joint.
	2. Performance requirements for operating windows/doors vs. non-operating components (§ 5.4)	2. Same performance for both.
	3. Design displacements (§ 5.5 & 7.2.5)	3. 0.010 x the greater adjacent story height.
	4. Orientation of horizontal displacement criteria (§5.5)	4. In the plane of primary mockup elevations.
	5. Weather air and water testing is performed subsequent to this test and weather these tests shall have pass/fail criteria or are for information only. Additionally the extent of repairs allowed if any, to wall prior to air and water testing. (§ 5.6)	
	6. Time duration of each cycle. (§7.2.3)	6. As determined by test agency
	7. Additional cycles and/or displacement direction. (§ 7.2.4)	
	8. Additional displacement tests of magnitude greater than design displacement test. (§7.4)	
	9. Pass/fail criteria for (§ 11.1): -Glass breakage/fallout -Post design performance	9. As outlined in test method (11.0) for building occupancy type
AAMA 501.5 –98 Thermal	1. Area of thermal testing (§ 5.2)	1. Limits of testing agency's chamber
	2. Extreme design temperatures including exterior surface temperature (§ 1.2)	2. Exterior high temperature 82 degrees C. (180 degrees F) Exterior low temperature –18 degrees C (0 degrees F) Inside temperature 24 degrees C (75 degrees F) (§ 8.4)
	3. Number of cycles (§ 1.3)	3. Three cycles
	4. Whether air and water testing is performed subsequent to this test. (§ 9)	

Notes: 1. Referenced section numbers are shown in parenthesis.
2. No default is provided unless shown.

6. Test 9 - E330 (Structural Performance) to ± 100% of design load and then to ± 150%. Individual structural elements of the wall system are typically analyzed using common engineering practices, however the overall reaction of the entire

system is most reliably determined by full scale testing. The structural overload testing of the wall system is used to check that the wall system has an adequate factor of safety beyond the design load against a system failure. At overload conditions, serviceability requirements, such as water leakage resistance should no longer apply. The structural test performed to overload condition is typically performed last since a structural failure during this testing would likely damage the mockup, rendering it unfit for subsequent testing.

7. Test 10 – ASTM E 331 Water penetration is rerun to check the performance of the wall system after it has undergone full design wind loading in the previous test.
8. Test 11 – ASTM E 330 Structural Performance testing is rerun at 150% of the design load to check the performance of the wall system against design overload.

ASTM E 2099 - Section 6. Construction

A specialty subcontractor commonly performs the role of constructing the exterior wall. Often, due to the subcontractor's expertise, they will also act in a design capacity to detail the exterior wall based on the specifier's performance requirements. Nonetheless, the E 2099 standard refers to this party as the builder.

One of the key elements in communication of the proposed exterior wall system is the preparation of shop drawings. For the mockup, the E 2099 standard requires the builder to prepare an explicit set of mockup shop drawings. These shop drawings are necessary to; 1) explain the proposed system to the specifier; allow the test agency to prepare a test chamber for the mockup; 2) show the agreed upon design of the mockup; 3) facilitate the test agency in checking that the installed system meets the design; 4) allow assistance in determining a solution if the mockup fails the tests, and 5) allow the recording of modifications to the wall system for future reference.

A review of the shop drawings by the specifier will provide the opportunity to verify the builder's interpretation of the performance specifications. This review will usually simplify the review of the project shop drawings subsequent to the mockup testing process. The test agency will design the test chamber based on the shop drawing dimensions and will provide the necessary structural support to attach the mockup. During construction of the mockup the builder can assess the shop drawing details and modify details for improved constructability. These modifications should then be recorded on as-built mockup shop drawings. It is important for all members of the project team to understand changes made to the proposed system and their affect on the construction and performance of the wall system.

The standard also recommends that the mockup be constructed by the builders personnel that will be responsible for erection of the wall system for the project. This allows the builders personnel to become familiar with systems, which are custom designs and begin to develop the methods of construction of the wall system for the project.

The time length for mockup construction is dependent on several factors, including: complexity of the wall system, material availability, and modifications made during the mockup process. The E 2099 standard recommends that the specifier observe some or all of the mockup construction. This allows the specifier to verify that the mockup is constructed in accordance with the shop drawings and at the same time gain a greater familiarity with the details of the proposed wall system.

ASTM E 2099 - Section 7. Testing

Section 7, the testing section of the E 2099 standard is directed to the test agency. Recognizing that the test agency is probably the most knowledgeable party in the mockup process, guidance directed to the test agency can be concise. Section 7 guidelines may also be educational to the other parties involved in the mockup testing process, in order to understand the test agency's role. Indeed, in some cases, the test agency needs to remind the other parties that the test agency is not the specifier and that the test agency's obligations are normally limited to those described in Section 7.

ASTM E 2099 - Section 8. Documentation

An important area where the E 2099 standard provides guidance is on mockup documentation. Documentation, as discussed in the E 2099 standard's statement of purpose, is a key link between the lessons learned in the mockup testing and the actual building construction. Cases abound of buildings that have had mockup tests, yet suffer facade performance problems. Investigations into these problems have often found that the lessons learned about problems in the mockup were not carried through to the constructed facade.

Successful projects responded to problems exposed during the mockup testing, first by trying out repairs on the mockup, and then when effective repairs are determined, implementing these same repairs throughout the construction of the façade. Figure 2 illustrates diagrammatically this process.

Conclusion

Building a better wall system can be accomplished by integration of the new standard E 2099 into the design, construction, and testing program for exterior wall system laboratory mockups. For over 40 years, laboratory mockups have proven themselves time and again as key components in the assurance of quality construction. Mockups are invaluable tools to learn lessons before construction begins, how complex and costly exterior wall systems perform. A detailed mockup test regime can not only verify performance requirements, but can also reveal overlooked faults in the exterior wall system. These faults can then be corrected on the mockup and re-tested to provide a quality check of the system. Compare this process to one where no mockup is constructed; no lessons learned before construction; and faults not exposed until after the building is complete. The cost for correcting these faults could be many times greater than what the mockup testing cost would have been.

This paper describes the manner in which the E 2099 standard is a means to improve the mockup process. The E6.55 committee members were charged with refining the mockup process. This process had developed over time from three relatively simple

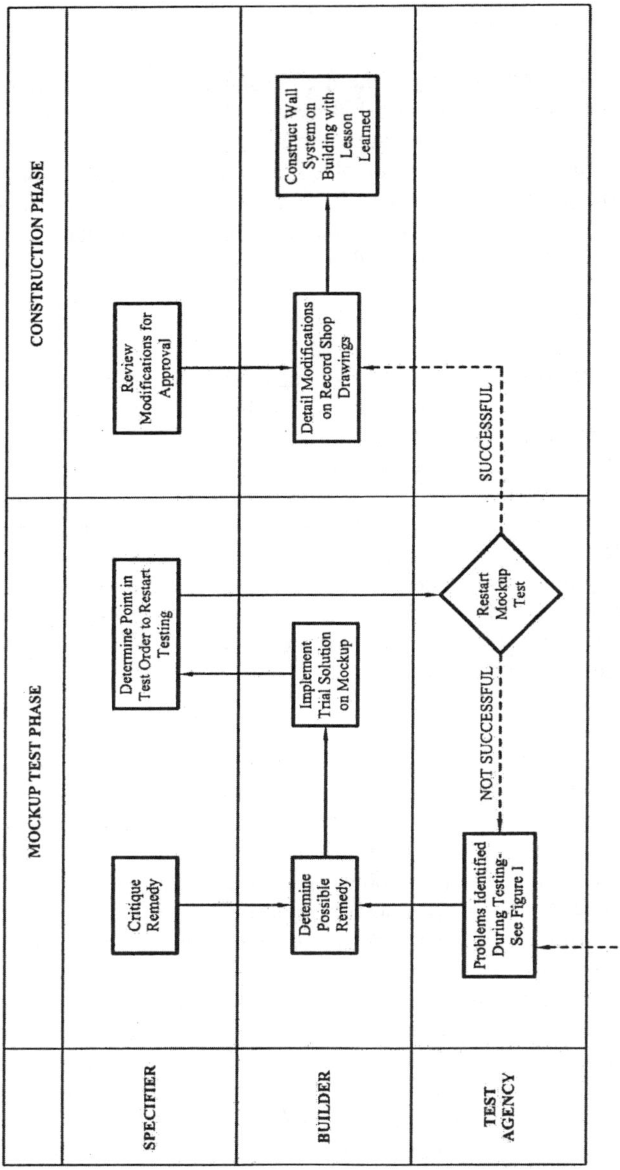

Figure 2 --*Flowchart of resolution of mockup problems*

tests to over ten current test alternatives, each with its own particular requirements. It falls mainly on the specifier to assure that each test and its particular requirements are specified correctly. In the absence of any stated requirements, default requirements are often used, which may not meet the specifier's desires. Therefore, well-prepared specifications are very important to the success of the test program.

The builder and the test agency also must interact to create a successful test, and a successful project. Most critical, is the involvement of the builder, before, during and after the test. Mockup shop drawings are essential. The mockup should be built by the same people entrusted with building the exterior wall system. The lessons learned during the mockup process should be documented in follow-up shop drawings and reports. When all these steps are accomplished the final result should indeed be a better wall, one which is capable of providing the level of performance expected of it.

References

[1] Kaskel, B. S., Scheffler, M. J., and Chin I. R., "Critical Review of Curtain Wall Mockup Testing for Water Penetration," *Water Leakage Through Building Facades, ASTM STP 1314*, R.J. Kudder and J.L. Erdly, Eds., ASTM International, West Conshohocken, PA, 1998.

[2] Johnson, P. G., "Building Exterior Wall Water Infiltration Control Using Quality Assurance Programs," *Water Leakage Through Building Facades, ASTM STP 1314*, R. J. Kudder and J. L. Erdly, Eds., ASTM International, West Conshohocken, PA, 1998.

[3] Hunt, William Dudley Jr., AIA, *The Contemporary Curtain Wall; Its Design, Fabrication, and Erection*, F. W. Dodge Corporation, New York, 1958, pp. 390–391.

[4] Sakhnovsky, A. A., "Full-Scale Performance Testing of Curtain Walls," *Exterior Wall System: Glass and Concrete Technology, Design and Construction*, ASTM STP 1034, B. Donaldson, Ed., ASTM International, Philadelphia, 1991, pp. 47–58.

[5] AAMA 501.1-94, "Standard Test Method for Metal Curtain Walls for Water Penetration Using Dynamic Pressure," American Architectural Manufacturers Association, Schaumburg, 1994.

[6] AAMA 1503.1-88, "Voluntary Test Method for Thermal Transmittance and Condensation Resistance of Windows, Doors, and Glazed Wall Sections," American Architectural Manufacturers Association, Schaumburg, 1988.

[7] AAMA 501-94, "Methods of Test for Exterior Walls," American Architectural Manufacturers Association, Schaumburg, 1994.

[8] AAMA Curtain Wall Design Guide, (CW-DG-1), American Architectural Manufacturers Association, Palatine, 1996.

Robert Bateman[1]

A Detailing Method for Improving Leakage Prevention of Exterior Wall Weatherproofing

Reference: Bateman, R., "**A Detailing Method for Improving Leakage Prevention of Exterior Wall Weatherproofing,**" *Performance of Exterior Building Walls, ASTM STP 1422*, P. G. Johnson, Ed., ASTM International, West Conshohocken, PA, 2003.

Abstract: As generally or often practiced, building design and construction detailing of weatherproofing exterior building walls relies on two-dimensional (2-D) graphic representations of building components to indicate materials used for weatherproofing. However, these details which serve a variety of purposes for different users do not show all the relevant materials needed to construct a weatherproof exterior wall. The various users of construction details, e.g., contractors, builders and material installers, utilize these 2-D drawings as visual instruction guides for constructing and installing building components. Common construction details are intended to show various components that serve a variety of purposes, such as structural support, fire-resistance, energy control, acoustic control, aesthetic appearance and incidentally weatherproofing. Different contractors during construction will review the same detail to obtain information relevant to the purposes of their particular trade. Construction details usually identify a material or component by name and not by the trade that constructs it or the purpose it serves.

Construction details, in addition to traditional uses, need to focus and emphasize weatherproofing materials and assemblies. Three-dimensional details should be used to show the junctures of exterior wall construction where weatherproofing installation is critical. Sequence (Step-by-Step) views should be used to show the layering of weather barriers and flashings in the proper order of installation. Enlarged cross-section details should be used to identify and show what weatherproofing materials are required and where they are installed.

Keywords: building construction, details, weatherproofing, flashing, exterior walls, leakage prevention, design

This paper is based on the premise that leakage of liquid water through the building envelope can be a consequence of poor building design and poor or absent construction details [*1*]. The weaknesses of the current practice of construction detailing are outlined. Using example details from a case study, an improved method of weatherproofing detailing for the design phase and construction phase is presented for consideration by architects, building designers, waterproofing consultants, contractors and product manufacturers.

---

[1]Senior Staff Architect, Simpson Gumpertz & Heger, Inc., 222 Sutter Street, Suite 300, San Francisco, CA, 94108.

Problems with Current Detailing Practice

Incomplete Information Shown

Current 2-D architectural details incompletely show the relevant information that is required to build a complete exterior wall weatherproofing system.

Exterior Wall Weatherproofing is Not Considered as a System – Conventional architectural practice does not sufficiently recognize weatherproofing as a discipline or system that needs to be graphically detailed with a consistent and continuous assembly of water control barriers, seals, flashings and terminations. In current practice, most architectural details for exterior wall components do not graphically represent all the components of the weatherproofing system, especially flashings and weather barriers.

Lack of a Single Trade Responsibility for Exterior Wall Weatherproofing – Weatherproofing details should show a continuous system of components that can be constructed in a coordinated and consistent manner. With current building practice, the weatherproofing of an exterior wall is not under the responsibility of a single trade, although the general contractor is responsible for overall coordination and sequencing of work. Most wall weatherproofing installations require several trades to assemble and construct various weather barriers and flashings for windows, doors and wall claddings. If a detail used for construction is incomplete, then the separate trades installing parts of the weatherproofing assembly may not identify and add the necessary materials missing from the detail. And unlike current roofing practice, where the roofing contractor is responsible to integrate the different parts of the roof assembly, there is no corresponding trade with the comprehensive knowledge needed for integrating all the parts of most exterior wall weatherproofing.

Limitations of Two-Dimensional (2-D) Details – Conventional construction details and weatherproofing product manufacturers' details are predominately represented by two-dimensional, pictorial details. Information about the real world is limited to what can be included in two-dimensional representations of objects. The length, width and depth of an object cannot be shown from one drawing viewpoint using the standard drawing presentation views of plan, elevation or section; multiple views are required [*2*]. Typical or common building conditions whether the views shown are plan view, elevation view or cross-section view are more easily conceived in the designer's mind and more easily drawn to scale in 2-D details. However, 2-D details often do not show all the critical weatherproofing elements required because of the limitation of illustrating the junctures and terminations of multiple materials.

Detail Scale Is Too Small – Buildings and their wall assemblies are traditionally shown as 2-D drawings as plan views, elevation views or cross-section views. Conventional construction details are drawn at a scale too small to distinguish the layers of weather proofing – usually represented with lines drawn too thin to be distinguished from other adjacent materials. Most construction details are drawn at the smallest practical scale as judged by the designer. The intent of most architectural details should be to show the relationship of the important parts of an assembly to each other in scale. However, what can be an adequate scale for identifying most exterior wall components is often too small for identifying thin sheets of weatherproofing materials, such as, flashings and weather barriers.

Limitations of Cross-Section Views – Typical Cross-section details are usually insufficient and incomplete in showing the junctures of materials and components at terminations, transitions and penetrations. Exterior wall details show the outline of building parts as if the exterior wall were cut open after assembly. The parts of the wall assembly are shown tightly together as they would be when built. Unfortunately, the weatherproofing components, such as seals and layers of weather barrier sheet materials are so thin or small that any drawing scale less than one-half size does not adequately differentiate the separate materials clearly. Flashings, weather-resistant barriers of sheet materials, are actually only as thick as a few mils to perhaps 1/16 inch (0.03 mm to 1.59 mm). The typical detail is drawn as an assembly of many layered or lapped materials shown as parallel lines making it difficult to visually distinguish separate components. These components are graphically represented in thin lines that remain a visual blur if drawn next to each other or against the outline of another object in the typical detail.

Description by Written Specification Only

Many project drawings with notes; manufacturers' recommendations for product installation and industry references describe material installation without providing any form of illustration. Written installation instructions can often be stated unclearly and subsequently be misunderstood. Written instructions can be found to result in a surprising number of varying interpretations when field inspections occur.

Worse than no pictures is too few words. Reviewers of building plans, too often, so as to become a cliché, encounter notes on details, such as, " FLASHINGS BY OTHERS" or, "WATERPROOFING PER CODE" and "INSTALL PER MANUFACTURERS RECOMMENDATIONS". These notes usually provide inadequate guidance for even knowledgeable contractors, and certainly not for unskilled tradespersons that often have the responsibility for installing most weatherproofing components on a building [*3*].

Lack of Weatherproofing Details from Industry References

The construction industry lacks references on weatherproofing details that can be used as a sole and complete reference for design or construction [*4*]. Industry references that do exist are often too general to be applicable without clarification for use with any particular project. Most industry references, which cover specific weatherproofing applications, may supply two or more options that must be selected for a specific project. However, the project specifications for construction rarely select the particular option indicated by a general industry reference when that condition has not been specifically detailed. Many industry references only provide basic or typical details (even if illustrated in 3-D) and do not show the termination, transition or penetration conditions encountered with conventional building practice. Unusual or unique details are presented by the industry to promote the possibilities of a particular material for unusual circumstances. Consequently, the details for a particular project most often have to be developed as unique drawings, if done at all.

Improved Method of Weatherproofing Detailing During the Design Phase

Communicating the relevant information needed for weatherproofing details requires some changes to what has been standard practice, since many of the current construction drawing and detailing methods are not effective.

This improved method of detailing uses, as a case study, an exterior wall recladding of a church structure in Concord, California. The building had experienced severe water intrusion through design and construction defects of the exterior wall weatherproofing. The existing exterior wall was constructed with wood framing, plywood sheathing, building paper, expanded metal lath and cement plaster (stucco). The stucco, lath, stucco accessories, building paper, and flashings were removed and replaced with the doors and windows remaining in order to correct weatherproofing deficiencies. The repair details presented in this paper focus on the weatherproofing components of the repair. The computer-aided design (CAD) drawings for the case study were developed during the project's design phase responding to as-built conditions known to require repair.

In order to address the lack of consideration given to weatherproofing as a system of interrelated components, a drawing or series of drawings are needed to provide an overview or context from which details can be referenced as needed. Like a roof plan or building section is the context for referencing roof details, the exterior wall elevations or building section typically serve for exterior wall detail references. Isometric views of the exterior building envelope can show the junctures of buildings that elevations and section do not show. Isometric building views can provide a better reference for exterior wall details. Wall weatherproofing components are most problematic to design and construct at building junctures, material terminations, transitions and penetrations. These areas are best shown with three-dimensional views.

Large-scale details showing cross-section views can be useful if the separate weatherproofing components are clearly shown.

Sequential details showing the step-by-step assembly of weatherproofing components easily illustrate the proper layering and integration of sheet materials that can not be otherwise understood with conventional cross-section details. Drawing building junctures with sequence views is a method to analyze the proper layering of sheet materials.

Three-dimensional (3-D) drawings have the advantage of being an analysis tool and providing a format for presenting construction details when the same drawings can be used to develop solutions for weatherproofing. As the detail is being drawn, the designer can consider options of materials, changes in sizes and alternate placement. Communication between the designer and the builder is improved using 3-D details of building junctures that show the conditions in the field that have to be addressed and allow a graphic evaluation of the constructibility of proposed solutions. Developing details showing views in sequence and in 3-D can take advantage of the use of CAD drafting. CAD can be used to set up the base drawings and develop the subsequent views. Creating the overall building design in 3-D with CAD models can provide a reference or context view of the building and a means to identify and reference exterior wall areas needing to be detailed.

Context View

Reference drawings are necessary to view the overall context. Reference drawings are used like a key map to show from where the details are taken and enlarged. A reference drawing, such as, an exterior wall elevation, building or wall section can be used as the drawing for the context view (Figure 1). These drawings can be used to identify junctures for detailing.

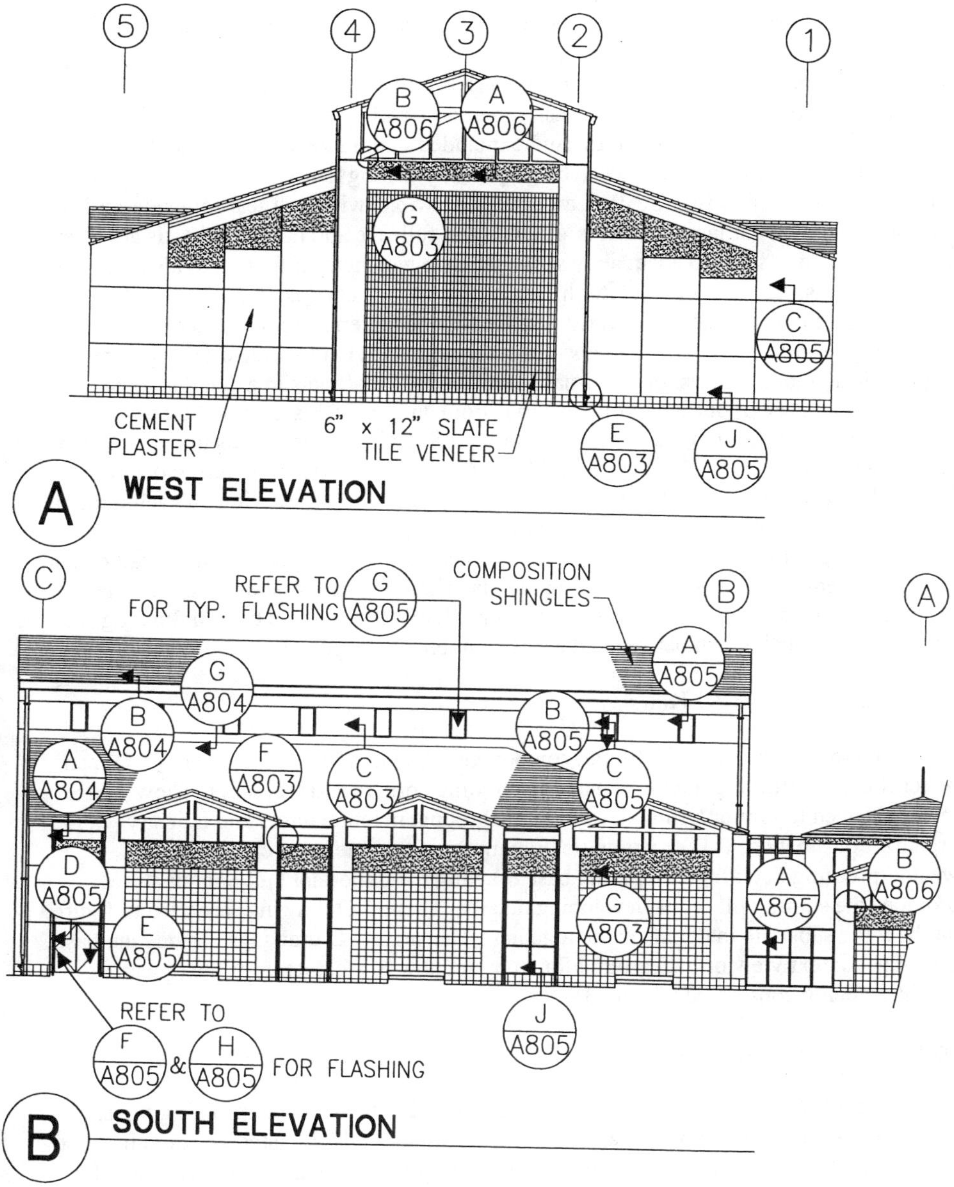

Figure 1 – *Context (Reference) View*

Enlarged Details

Cross-section details are still essential in communicating information about weatherproofing components. Large-scale cross-section details that can clearly distinguish components, especially sheet materials make for a better communication tool. Details presented at a large scale are easier to understand by contractors during construction. Multiple parallel lines can be included with large-scale details. Lines that may represent different materials can be graphically distinguished by their placement in the correct relationship (Figure 2). Large-scale drawings will be at least one quarter life size or 3" = 1' – 0" or larger. Even larger scales at full size and half size details are often the most effective scale to adequately show separate weatherproofing materials. Small-scale drawings, less than one quarter life size, by contrast, can lose the features of separate parallel lines that sufficiently distinguish one material from another.

Separation of Material Layers – In both large and small –scale drawings it helps to provide a graphic separation or visual space between materials in a detail to clearly show the boundary or outline of one material from another. This is important when layers of material occur lapped one over another. Showing the materials in contact with each other, as the actual condition would be built results in a visually blurred detail. Materials shown in the detail separated graphically provides a clear view of each material and its relation to other materials (Figure 2).

Exaggerated Line Thickness – The weatherproofing components in a detail can be graphically emphasized by exaggerating the material line thickness for sheet materials, such as, flashings and weather-resistant barriers. Differences in the type of drawing line can be used to distinguish between adjacent components (Figure 2).

Three-Dimensional (3-D) Views

Pictorial drawings of building components shown in three dimensions (3-D) can better illustrate building junctures, such as building corners and recesses to show where materials need to join or terminate [5]. Some product manufacturers and industry organizations that use 3-D or isometric drawings can provide good examples for weatherproofing detailing. Preparing base drawings of important building junctures provides the starting place for developing details. These 3-D drawings are used to figure out how the substrate materials come together. The same 3-D drawings provide the base drawing, context view or background for studying how to weatherproof the juncture or condition being considered (Figure 3).

Sequence (Step-by-Step) Views

Details that show the construction sequence of critical weatherproofing materials in a step-by-step manner provide a clear understanding of which materials come first and those that follow. The process of developing the sequence drawings acts as a check for constructibility. Materials that are out of place are readily identified (Figure 4).

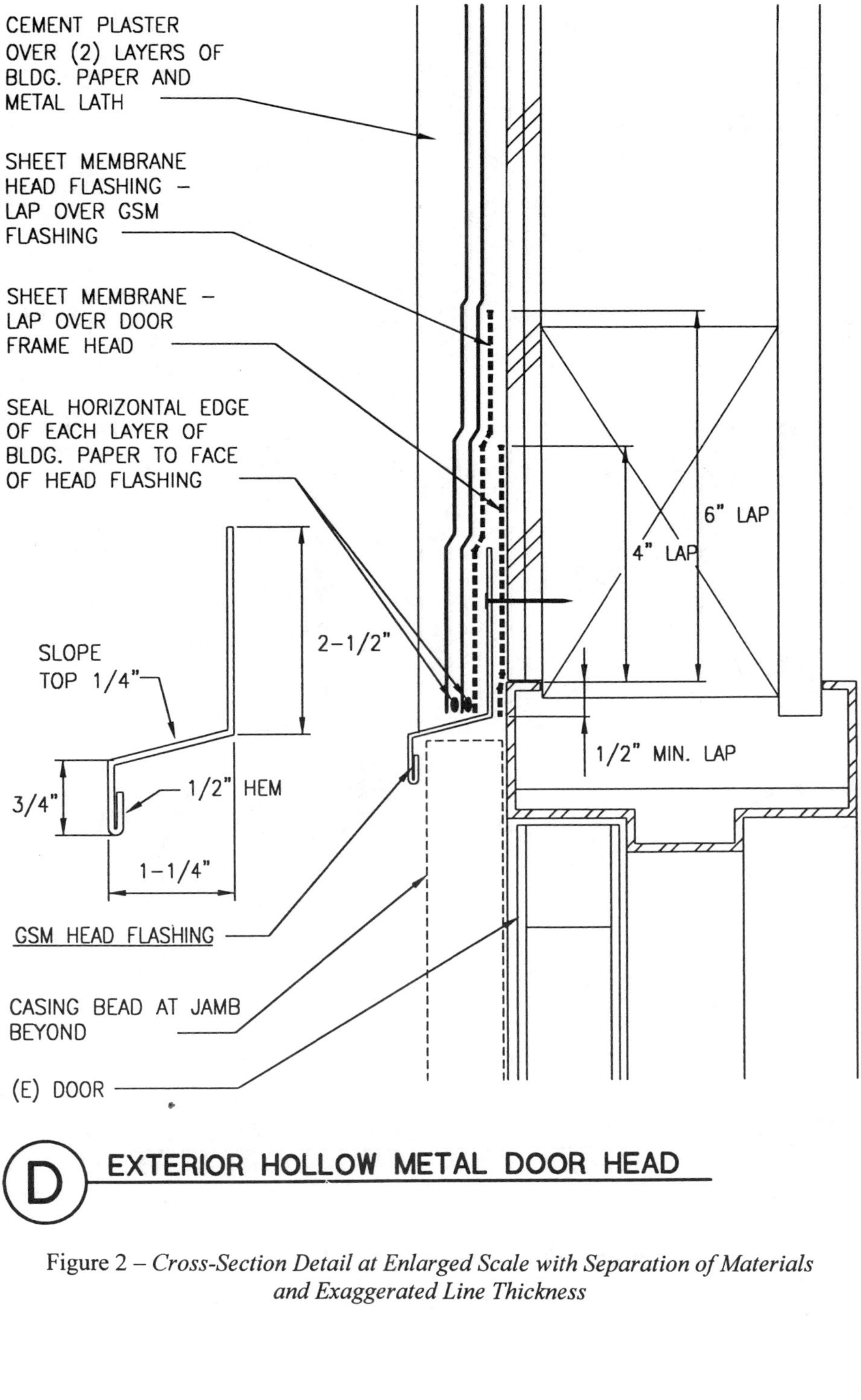

Figure 2 – *Cross-Section Detail at Enlarged Scale with Separation of Materials and Exaggerated Line Thickness*

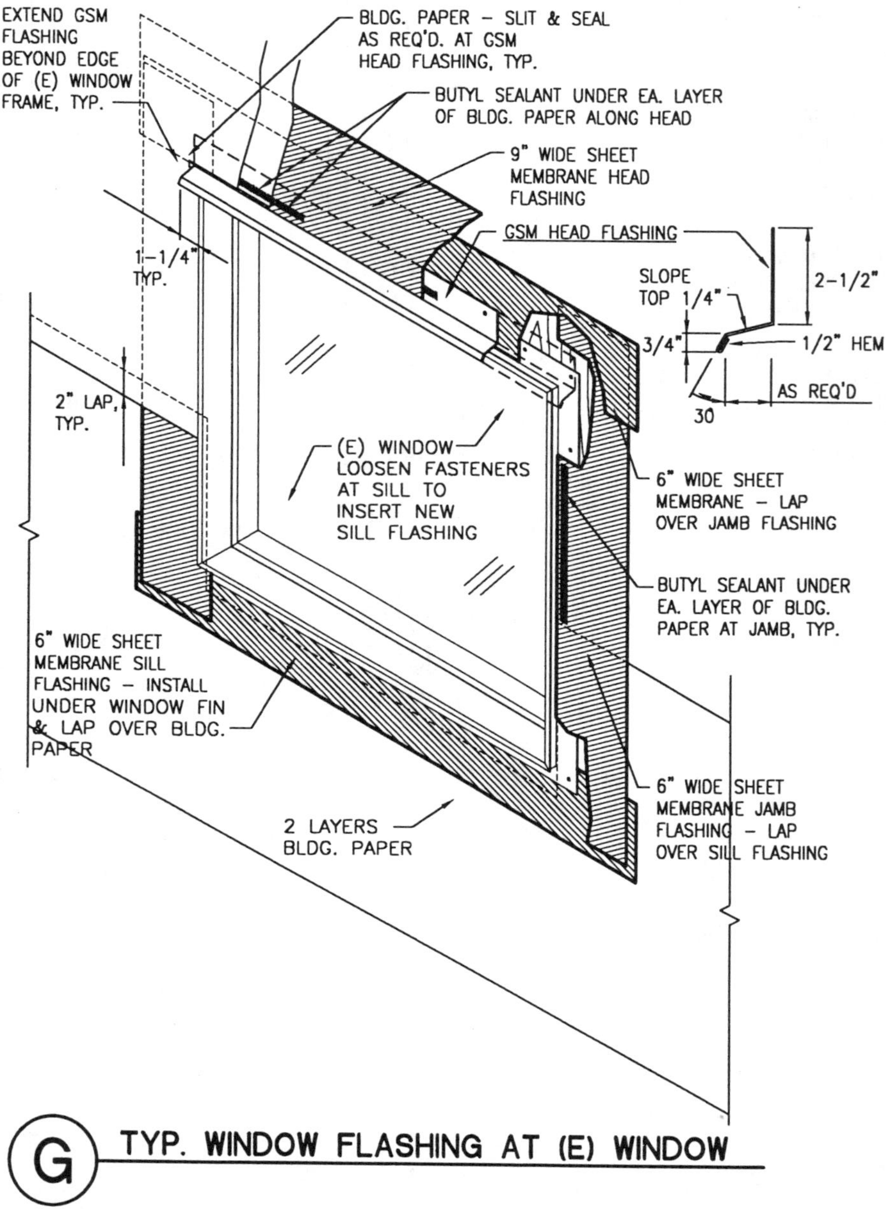

Figure 3 – *Three-Dimensional (3-D) Views*

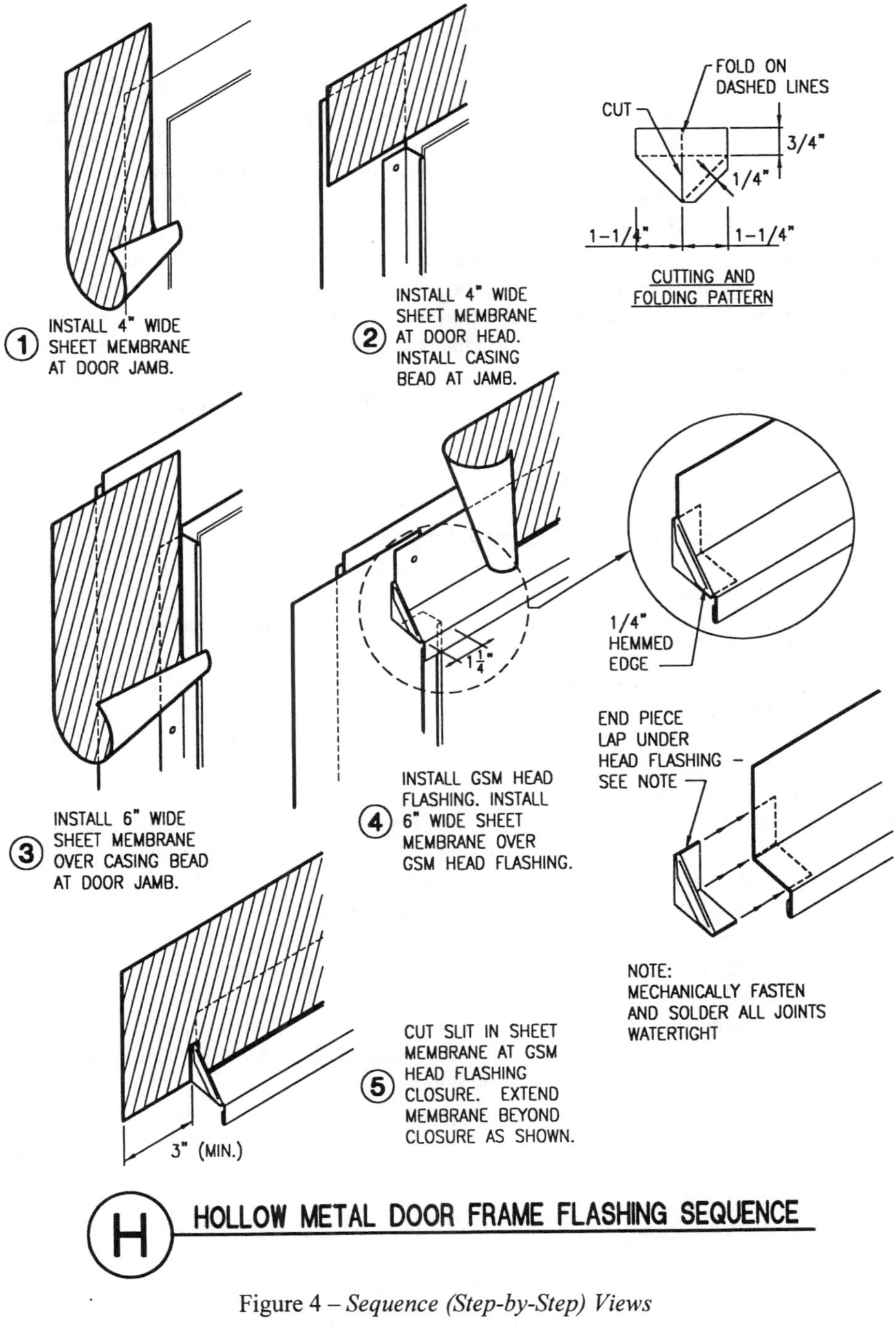

Figure 4 – *Sequence (Step-by-Step) Views*

Improved Method of Detailing during the Construction Phase

Detailing of the weatherproofing system can continue or be completed during the construction phase if this activity is planned for. The field participation of the designer responsible for the weatherproofing system can serve as the quality control activity for checking the appropriateness of the construction details. In many cases, with existing buildings, weatherproofing details can not be fully determined until the as-built conditions are revealed. With new construction, when the weatherproofing options have not been completely selected, this method can be used during the construction administration phase as the contractor requests clarifications from the designer.

The freehand sketches prepared during the construction phase of the case study presented here were anticipated for repairs to concealed defects and the necessary construction funds were budgeted. The owner, contractor and consultant decided that certain repair details would best be developed during construction to respond accurately and efficiently to as-built conditions.

Preconstruction Meetings / Construction Visits

Coordination of the overall building design and weatherproofing details is improved when the designer and general contractor review the weatherproofing details on the job prior to and during construction. Cross-section and 3-D details developed during the construction and at field visits can be prepared by the designer or contractor as free-hand sketches when efficiency and timeliness of communication is critical [*6*]. Freehand sketching was a skill once considered essential for students of architecture [*7*] to understand how buildings were assembled.

General Contractor's Superintendent – When the weatherproofing designer has the opportunity to meet with the contractor, the overall intent of weatherproofing system design and the identity and emphasis of critical elements of the details can be communicated (Figure 5). With larger projects, that support the contractor using an on-site superintendent, there is opportunity to respond to the designer with discussion on the relative difficulty, cost impact and scheduling aspects of constructing the details. Consequently, the designer and contractor work together on refining the details leading to cost-effective and practical solutions. This can be a workable process whether the contract is negotiated or bid when the both the designer and contractor plan for the time and cost involved.

Subcontractors – When the contractor or construction superintendent arranges meetings with various trades responsible for the weatherproofing work, the designer can review the critical or problematic details and obtain feedback prior to installation. The superintendent can identify the various components of the work and trades that need to be coordinated during the progress of construction (Figure 6). In response to subcontractor feedback, the designer can revise details when required and forward them to the superintendent.

If shop drawings are prepared for the project, they should be presented with 3-D, sequence views and large-scale details when it is necessary to understand how the weatherproofing details will address terminations, transitions and penetrations.

Prototype Evaluation – Taking a detail to the job site or to a mock-up construction prototype provides the opportunity to verify that the detail can be installed as intended. Both designer and contractor can determine how constructible the detail will be. Changes can be discussed that may simplify or improve the detail.

Sequence Views – As with the design phase, sequence diagrams showing the step-by-step procedures of material applications are developed during the construction phase. These aid the designer to check the practicality and constructibility of the proposed design. They can be used as field guides for the trades' use in assembling the weatherproofing components, especially flashings that are usually built up in layers or lapped over one another in a necessary sequence (Figure 7). Sequence details also act as an aid during construction observation to verify that the assembly of materials is progressing properly.

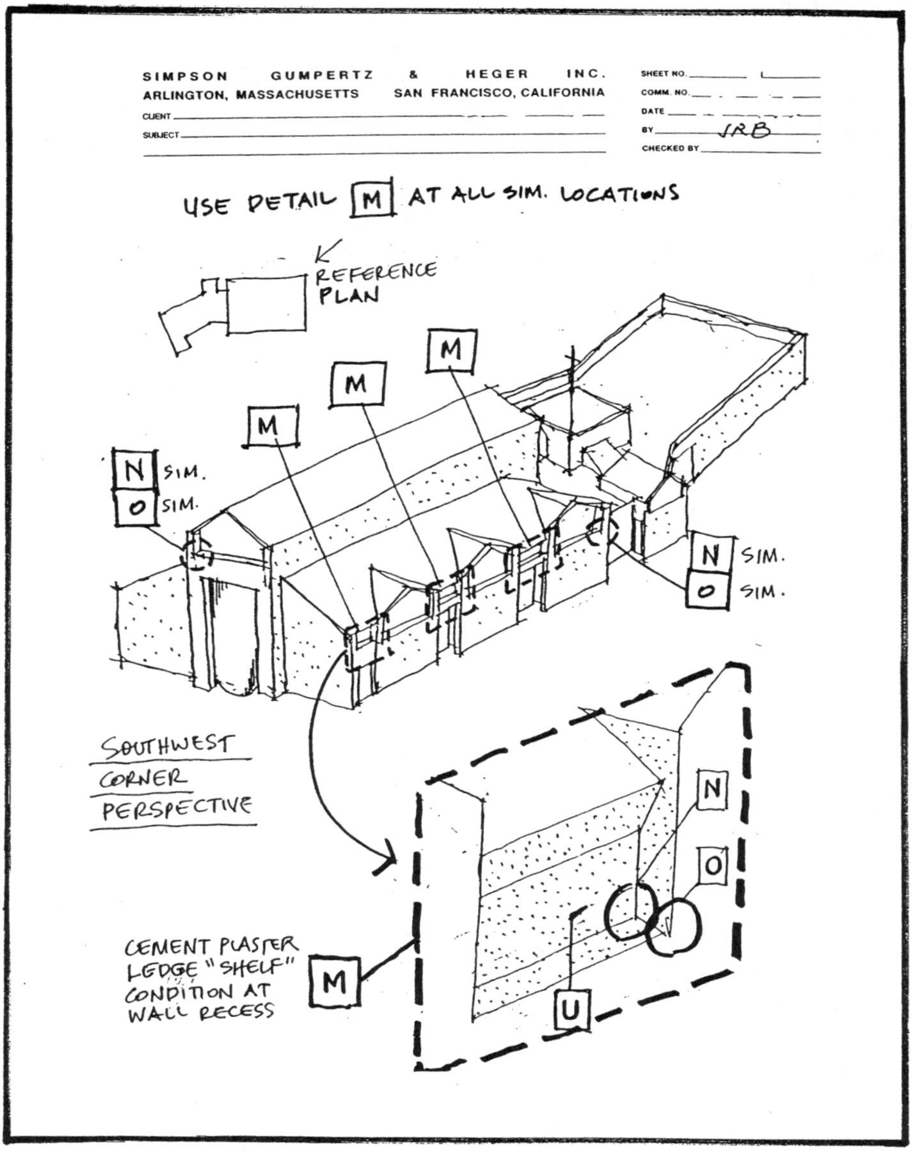

Figure 5 – *Context (Reference) View*
(freehand sketch)

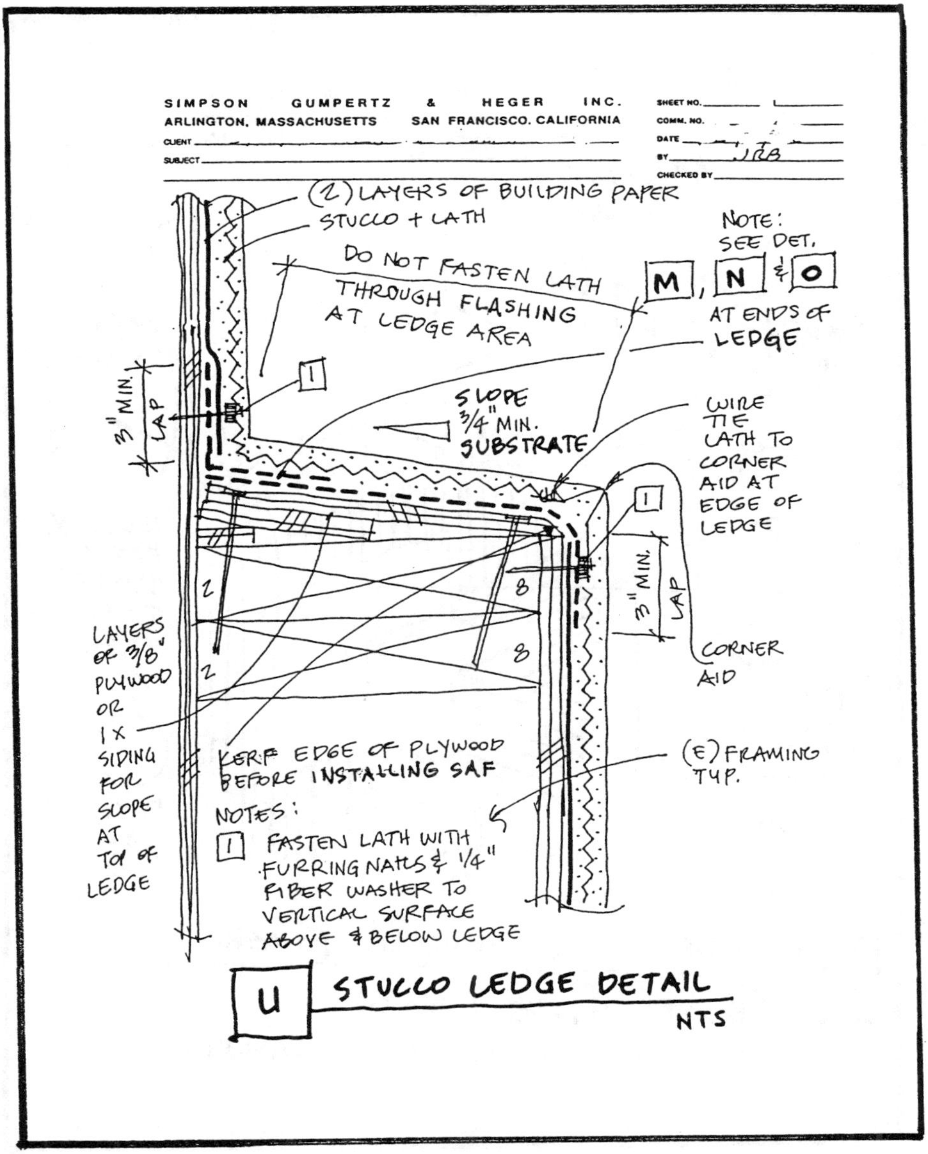

Figure 6 – *Enlarged Cross-Section Detail*
(freehand sketch)

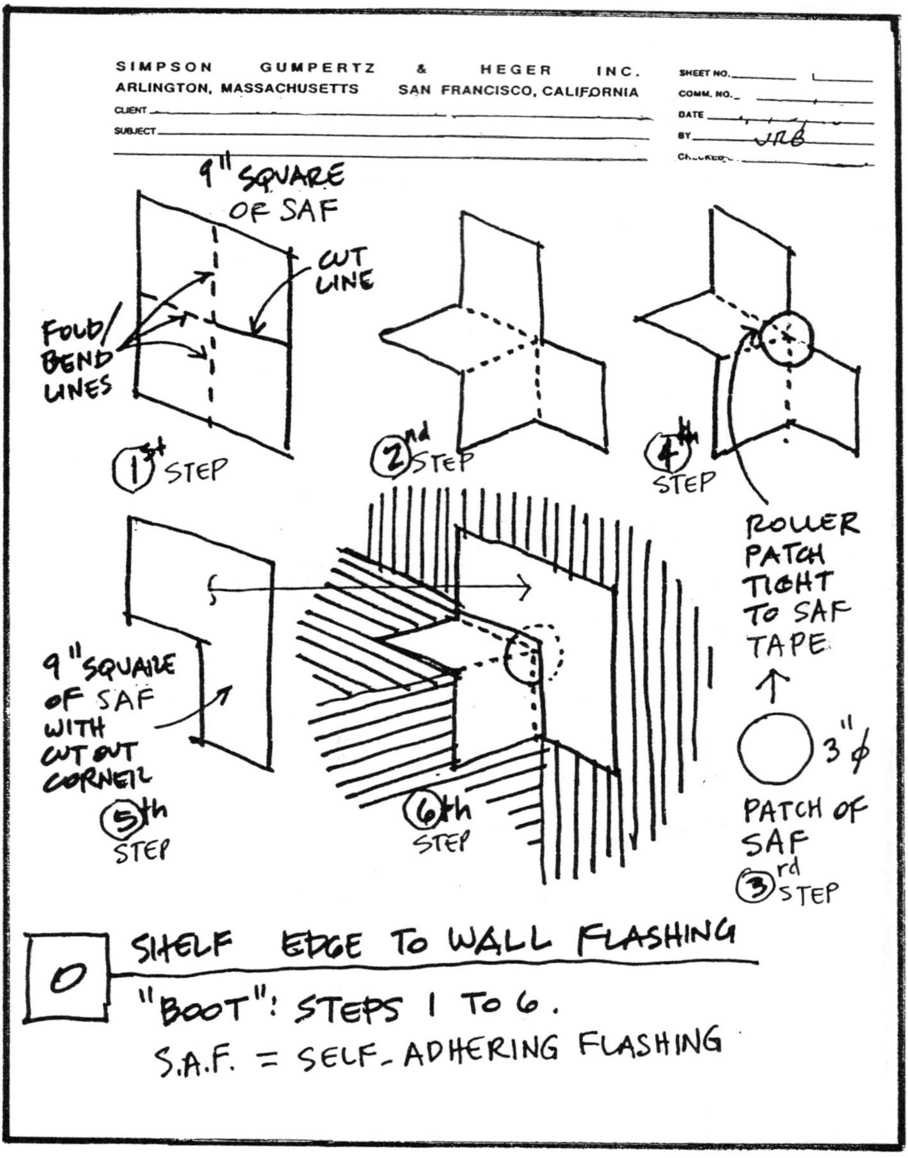

Figure 7 – *Combined 3-D & Sequence Diagrams*
(freehand sketches)

Conclusion

Conventional 2-D construction documents are inadequate when weatherproofing is not detailed as a comprehensive system of interrelated components. Misunderstandings occur when construction information is missing, misplaced or only confusing details are available. Construction based on poor details or construction based on guesses about information missing from details contributes to performance failures of exterior wall weatherproofing, especially at complex junctures and unique wall conditions.

The use of 3-D (isometric) details, sequence diagrams and enlarged cross-section details that emphasize weather barriers and flashings improve communication by providing clearer design documentation to builders. The following advantages result from the use of these improved detailing methods for exterior wall weatherproofing:

- Ease of Analysis – 3-D and sequence view drawings provide a detailing method useful to analyze difficult construction junctures.
- Ease of Detailing in Pictorial Format – 3-D details reduce the time needed to conventionally draft scaled drawings and provide a clearer format to visualize three-dimensional conditions.
- Ease of Communication – Isometric, pictorial details are readily understood by designer, general contractor, subcontractors and individual trades persons.
- Saves Time During Construction – Hand sketches developed for critical details in the field results in quicker, more practical solutions to weatherproofing.
- Ease of Follow-Up – Sequence view drawings provide an effective and efficient means of checking the as-built details during construction.

References

[1] Kubal, M. T., *Waterproofing the Building Envelope*, McGraw-Hill, Inc., New York, 1993, pp. xiii, 6.

[2] Lowndes, W. S. and Emerson, D. B., *Architects' Blueprints and Specifications*, International Textbook Co., Scranton, Pennsylvania, 1927, pp. I-6.

[3] Bateman, R., *Nail-On Windows: Installation and Flashing Procedures for Windows and Sliding Glass Doors*, DTA, Inc., Mill Valley, California, 1995, pp 6-27.

[4] Ibid.

[5] Bourne, F. A. and von Holst, H. V., *Architectural Drawing and Lettering*, Technical World Magazine, Chicago, Illinois, 1908, p.39.

[6] Sands, H., *Wall Systems: Analysis by Detail*, McGraw-Hill Book Co., New York, 1986, p.5

[7] Field, W. B., *Architectural Drawing*, McGraw-Hill Book Co., New York, 1922, p. 1

Kevin C. Day[1]

Connectivity of the Air Barrier and Building Envelope System: Materials, Process, and Quality Assurance

---

Reference: Day, K. C., **"Connectivity of the Air Barrier and Building Envelope System: Materials, Process, and Quality Assurance,"** *Performance of Exterior Building Walls, ASTM STP 1422,* P. G. Johnson, Ed., ASTM International, West Conshohocken, PA, 2003.

Abstract: The engineering of air barrier and building envelope details for the construction of exterior walls requires that construction sequence and quality assurance be addressed. Further, the assembly of specific materials should not be based merely on "what was used in the last project" since the interior and exterior conditions of a given building may be vastly different from the proverbial last project.

The word "connectivity" is used in the information technology industry, and this word has been deliberately selected for the title of this paper to describe the inter-connection of materials that form an air barrier system in exterior wall assemblies.

This paper considers the air barrier system, as it relates to moisture management strategies for condensation and precipitation control, exterior and interior conditions, and the quality assurance of the air barrier installation. An example wall assembly is presented to demonstrate some of the issues that must be considered during the construction sequence.

Keywords: air barrier, air leakage, condensation, connectivity, heat loss, moisture management, quality assurance, rain penetration, wall assembly

Introduction

The air barrier is often referred to as a specific material in a wall assembly; however, it must be understood that the air barrier is in fact a continuity of materials in a building envelope. The word "connectivity" is used in the information technology industry to describe the interconnection, or linking, of computers through electronic networks. As the complexity of variables for computer networks can be terribly complicated to the neophyte, the prospect of designing an air barrier system can be equally complicated to a designer who does not consider the consequences of selecting inappropriate, and/or incomplete connections for controlling air leakage, i.e., making the air barrier system continuous. The operative word in the previous sentence is "*system*."

---

[1] Associate, Building Science Specialist, Morrison Hershfield Limited, 235 Yorkland Blvd., Suite 600, Toronto, ON, M2J 1T1.

The word "connectivity" has been deliberately selected for the title of this paper to describe the interconnection of materials that form an air barrier system; thereby creating an assembly that can adequately sustain the loads imposed, as part of the building system as a whole. The connections of the material components that form the air barrier system must be part of the building envelope design, since these connections can be achieved by a number of methods. Further, the ability of the various connections to perform adequately over the service life of the wall assembly must also be evaluated to ensure durable connections are constructed.

Purpose of the Air Barrier

As part of the preamble to the discussion presented in this paper, the control of air flow, water vapor, rain penetration, and thermal transfer were cited as being principle requirements of an exterior wall, as summarized by Hutcheon over 35 years ago [*1*]. Further in this regard, Brown, et. al. [*2*], delineate that an air barrier must be defined as a series of materials that are assembled to meet the following requirements:

- Constructed of air tight materials,
- Continuous through the building envelope,
- Strong enough to resist the air pressure loads, such that the air barrier transfers these loads to the building structure, and have enough rigidity or support so that deflection under load is accommodated in the specific wall design,
- Durable enough to provide the necessary performance in the service environment anticipated,
- Buildable.

The ultimate objectives of the air barrier are to prevent (or reduce the potential of) condensation, to maintain thermal comfort within the building, and reduce the energy consumption of the building system. The necessity for controlling air leakage in hot and cold climates are equally important towards these objectives.

When designing a wall assembly, it is important to consider what other properties the air barrier system may be expected to perform, such that water vapor, rain and/or heat flow may also be aspects of the design intent for the air barrier. The plane of air tightness, the vapor barrier, and drainage plane could all be separate, or be a combination thereof, depending on what materials are selected, and in turn, how these materials are connected. Consideration must be given to the inter-relationship of moisture management properties (or mechanisms) within the wall assembly, and how these can relate to, or be influenced by the proximity and continuity of the air barrier system. The purpose of the air barrier must be defined for a wall assembly to ensure that the continuity is achievable, and that the other principle requirements of a wall have been implemented into the design, as either part of the air system or some other elements of the wall assembly.

There are several references that designers may employ to design a wall assembly to prevent condensation due to air leakage and vapor diffusion .Two such resources that have been published by Handegord [*3*] and Lstiburek [*4*], and both provide basic design tools to determine the occurrence of the dew point temperature, the basis of determining the appropriate placement of insulation, and Lstiburek provides an useful tool to determine the potential paths of moisture movement through a wall assembly.

The following Tables 1 and 2 delineate the potential functions of an air barrier for a set of example wall assemblies, for both cold and hot climate scenarios. These tables are

Table 1 – *Example Wall Assemblies, Cold Climate, Functions of the Air Barrier*

	Cold Climate with Moderate Summers			
Assembly	Plane of Air Tightness	Dew Point Temperature	Vapor Barrier /Retarder	Drainage
Brick Veneer & Cavity Rigid Insulation Sprayed Bitumen Gypsum Sheathing Insulated Steel Frame Sheet Polyethylene Interior Gypsum	Composite formed by bitumen mastic spray, supported by sheathing.	Occurs within the steel stud cavity.	Polyethylene, dew point requires air barrier to be vapor permeable.	1. Brick Veneer 2. Sprayed Bitumen
EIFS Cladding over, Polymeric Cement Concrete Masonry Furring w/ Cavity Interior Gypsum	Polymeric Cement rendering onto block	Occurs within the EIFS.	Polymeric Cement	1. EIFS 2. Polymer Cement
Vinyl Siding Rigid Insulation House Wrap Wood Sheathing Insulated Wood Frame Sheet Polyethylene Interior Gypsum	Interior Gypsum	Occurs within the wood stud cavity.	Polyethylene, dew point requires air barrier to be vapor permeable	1. Vinyl 2. Housewrap

Table 2 – *Example Wall Assemblies, Hot Climate, Functions of the Air Barrier*

	Hot Climate with Very Hot & Humid Summers			
Assembly	Plane of Air Tightness	Dew Point Temperature	Vapor Retarder	Drainage
Brick Veneer & Cavity Bitumen Sheet Gypsum Sheathing Insulated Steel Frame Interior Gypsum	Sheet bitumen, supported by sheathing.	Occurs within the steel stud cavity.	Sheet bitumen, high permeance to the interior of this bitumen.	1. Brick Veneer 2. Bitumen Membrane
EIFS Cladding over, Polymeric Cement Concrete Masonry Furring w/ Cavity Interior Gypsum	Polymeric Cement rendering onto block	Occurs within the EIFS.	Low permeance elastomeric stucco finish coat.	1. EIFS 2. Polymer Cement
Vinyl Siding Building Paper Wood Sheathing Insulated Wood Frame Interior Gypsum	Interior Gypsum	Occurs within the wood stud cavity.	Building paper medium to high permeance material only	1. Vinyl Siding 2. Building Paper

provided only as a review of the principle requirements of the air barrier, relating to the moisture management properties of the given assembly.

Based on Tables 1 and 2, the variables such as indoor and outdoor design temperature and relative humidity would obviously affect, and possibly adversely alter, the performance of the assemblies that are given here as examples. The drainage properties are delineated as being primary (1) and secondary (2). Typically, the exterior cladding provides the primary deflection and/or moisture storage of precipitation. The secondary barrier provides drainage and/or resistance to moisture ingress when incidental moisture penetrates beyond the cladding. A crucial consideration in contrasting Tables 1 and 2 is the implication of devising an assembly of materials that may be suitable in one type of climatic environment, but not another. In hot, humid climates, it is not uncommon to find condensation problems, which may result in some building owners having the perception that there may be a rain penetration problem. In fact, warm moist exterior air may deposit water within the building assembly via air infiltration, and this can severe in hot, humid climates, where buildings are air conditioned, and mechanically de-pressurized [*5*].

When considering the moisture management capabilities of a given wall assembly, the ability of the wall to deflect rain penetration and drain incidental moisture should be verily understood. In Straube and Burnett's paper, "A Review of Rain Control and Design Strategies," [*6*] the specific mechanisms of moisture management are delineated as drainage, storage and transmission. The relationship of these properties to the design of a given wall assembly should be verily understood by a prudent designer.

The design of any air barrier system, and building envelope, should give consideration to the inter-relationship between the building mechanical systems, stack effect, and compartmentalization of occupied spaces [*7*]. These factors are beyond the scope of this paper, nevertheless, a prudent designer must be cognizant of the implication of these aspects of air flow within the building, as they relate to the air barrier system.

When a set of calculations for a wall assembly, based on realistic design extremes for a given project, determine that there is potential for condensation to occur, it is very important to consider how this potential moisture could affect the assembly. For example, insulated framed wall assemblies are typically at greater risk of condensation simply by the fact that there is a measurable temperature differential across this portion of the wall assembly. The main concern then becomes how much moisture could condense, and whether or not there may be an opportunity to vent this moisture out of the assembly. The cause of the condensation, and how much it may accumulate, will directly influence the damage functions associated with wood (rotting) and steel (corrosion) framed wall assemblies. If venting of this condensation is expected to occur, then it must obviously not interfere with the continuity of the air barrier system. Moderate to high amounts of moisture that may occur in framed wall assemblies should almost always be limited to venting outside of the structural elements, i.e., the sheathing of framed walls [*8*], and the sheathing should always be protected with an adequate moisture barrier.

Connectivity of the Air Barrier System

Manufacturers have typically employed the ASTM Test Method for Determining the Rate of Air Leakage Through Exterior Windows, Curtain Walls, and Doors Under Specified Pressure Differences Across the Specimen (E 283) to measure the air permeance of a given material. The measured values of air leakage through a specific contiguous material have been commonly much less than 0.10 $L/s/m^2$ as measured at

75 Pa of pressure. However, in the 1995 National Building Code of Canada, an air barrier material must have an air permeance less than 0.02 L/s/m^2, as measured at 75 Pa. Assuming that the materials have been adequately qualified by this criteria, the next priority is to design the physical connection of the materials. These connections must be assessed in some capacity, as indicated by Brown et. al., such that an air barrier system should provide a maximum air leakage between 0.05 – 0.20 L/s/m^2, measured at 75 Pa, depending on the water vapor permeance of the outermost non-vented layer [*9*]. Vapor permeance should be tested according to the ASTM Test Method for Water Vapor Transmission of Materials (E 96), and for building materials concealed within the wall assembly, it is preferable to utilize the water method.

It is necessary that the given air barrier system be able to demonstrate adequate resistance to high air pressure differentials, consistent with the anticipated wind loads on the given project. Structural wind load testing is normally performed in accordance with ASTM Test Method for Structural Performance of Exterior Windows, Curtain Walls, and Doors by Uniform Static Air Pressure Difference (E 330), including the measured sustained loading. Further to this test method, it is useful to have the specimen subjected to cyclic and gust wind loads to determine if excessive wind loads cause an deleterious effects to the air barrier system. After comprehensive wind load testing, it is crucial to measure the air leakage to determine whether or not the air barrier system was affected by the wind load testing, as per ASTM E 283.

It is prudent for a designer to create a list of the required details for the building envelope, giving particular attention to the interface details of the air barrier system. These details normally include the interfaces of the wall assembly with windows, penetrations, balconies, roofing, foundation waterproofing, etc. In each detail, one or two specific materials should be specified to create the physical connection of the air barrier system. The other functions of the air barrier must also be considered in this regard. Combining a mechanical and/or adhesive seal into a supplementary air seal is prudent. This ensures that if ever there is a defect in any isolated connection of the air barrier components, it is offset by the air seal of the adjoining material.

In each project, depending on the given scope of work, there are several procedures that can be easily employed to ensure satisfactory air barrier connectivity. Among the most important variables that will affect the performance of an air barrier system is the ability of the given air barrier connection to accommodate construction tolerances.

Connectivity of the Air Barrier Details

As an example to demonstrate the many conditions critical to connection of the air barrier system, an example of a wall system (which has been rehabilitated) is presented. Figure 1 is a detail of a window jamb interface with an air barrier system, applied onto a steel frame wall with sheathing, and brick veneer. In this wall assembly design, as it described in Table 1, the primary air barrier is the composite formed by a bitumen mastic spray-applied onto the glass fiber faced gypsum sheathing. This air barrier has been designed and detailed to provide a drainage plane for the brick veneer cladding.

Most of the materials selected to perform the function of an air barrier are fairly durable when being subjected to moisture, with the obvious exception of gypsum based materials. When self-adhering sheet and spray-applied membranes are installed over gypsum-based sheathing boards, the composite material can provide a durable air barrier, provided that the gypsum sheathing is structurally supported, and kept dry.

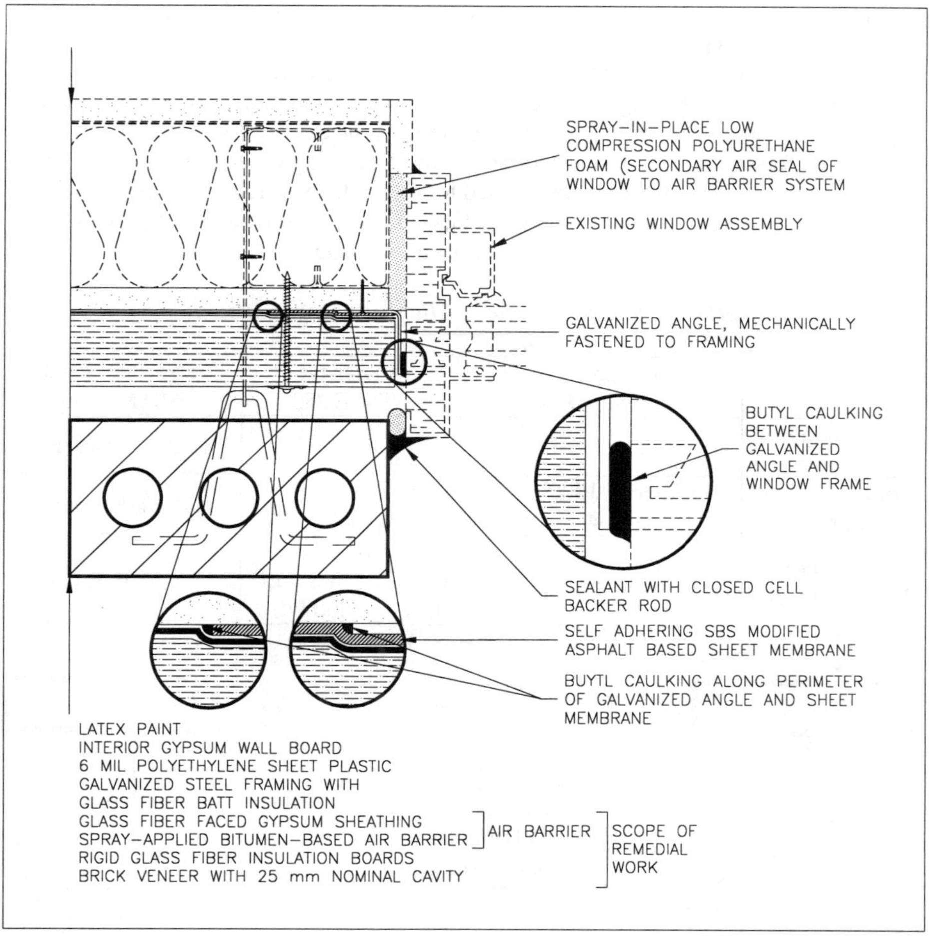

Figure 1 – *Window Jamb Detail*

This particular detail was developed for the retrofit of the exterior wall assembly, such that the brick veneer and sheathing were to be replaced. As part of the investigation, it was determined that the wall assembly did not have a satisfactory air barrier system, therefore, part of the remedial work included the installation of an air barrier. The interior gypsum finish, the steel framing, and the windows were all left in place. To achieve an effective air barrier connection, the perimeter of the windows were filled with sprayed-in-place polyurethane foam. Since the steel framed cavity was insulated, and the polyethylene vapor barrier was observed to be continuous, the air barrier system was required to be vapor permeable, and capable of being supported by a gypsum based sheathing. Further in this regard, it was determined that this air barrier system must have connected directly to the window frame. Therefore, the polyurethane foam provided a secondary air seal. A similar detail was developed for the sill detail. In this example, the existing condition of the window frames was satisfactory, such that the eventual replacement of the fenestration system would not be expected for at least another 20 – 30 years. The primary attachment of the air barrier system to the windows was achieved by installing galvanized steel angles at the perimeter of the windows and caulked with butyl.

The major components of the air barrier system for the example in Figure 1 are summarized for air leakage, vapor permeance, and structural properties as shown in Table 3. The bitumen based materials were selected for their inherent moisture resistance. In reviewing test reports and manufacturer's published technical literature, the properties determined by testing were representative of the design details.

Table 3 – *Properties Summary of Major Air Barrier Components*

Material	Wind Loading ASTM E 330	Air Leakage ASTM E 283	Vapor Permeance ASTM E 96
Sheathing Board	3350 Pa	< 0.01 L/s/m² @ 75 Pa	1320 ng/Pa•s•m²
Spray-Applied Bitumen	1 hour @ 1000 Pa 10 seconds @ 3000 Pa	0.054 L/s/m² @ 75 Pa	243 ng/Pa•s•m²
Self-Adhering Bitumen Sheets	1 hour @ 1000 Pa 10 seconds @ 3000 Pa	< 0.01 L/s/m² @ 75 Pa (after gust load of 3000 Pa)	2.8 ng/Pa•s•m²

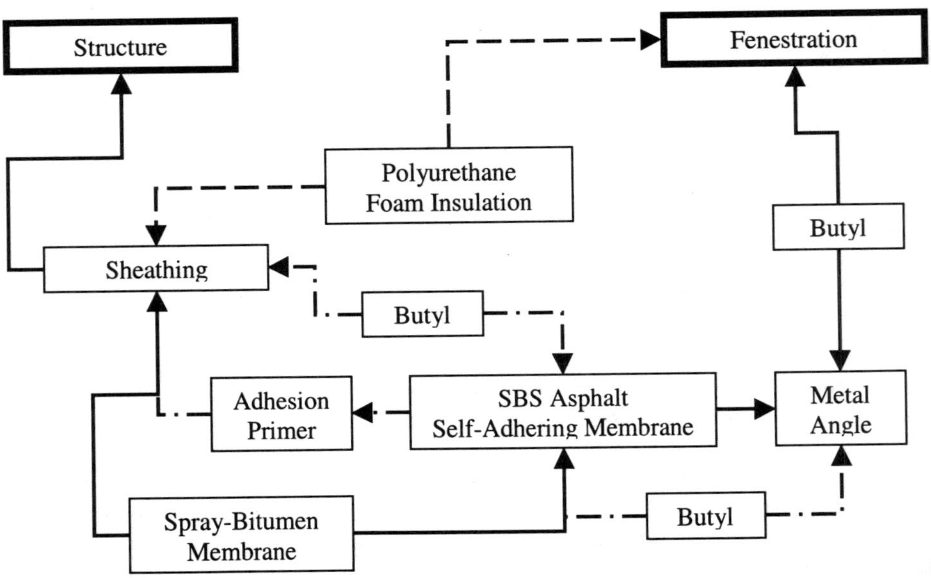

Figure 2 – *Material Connections from the Structure to the Window (as per Figure 1)*

The crucial aspects of the continuity of the air barrier detail illustrated in Figure 1 can be best summarized by the physical connections of the material components between the window frame and the wall assembly (in this case, the windows were judged to be reasonably air tight). In creating this detail, consideration was given to the following aspects:

- Construction Tolerance: the ability of the detail to adapt to the proximity of the window frame in relation to the plane of the sheathing, estimated in this example to be ± 12 mm, accommodated by the installation of a light gage galvanized steel angle mechanically fastened back to the steel framing.
- Redundant Connections: the physical connections of the primers, caulking, self-adhering and spray-in-place materials all rely significantly on the surface preparation (cleaning) and workmanship of the installation. By designing the connections to have redundant interfaces, this provides greater assurance of achieving an air tight transition between the window and wall assembly.

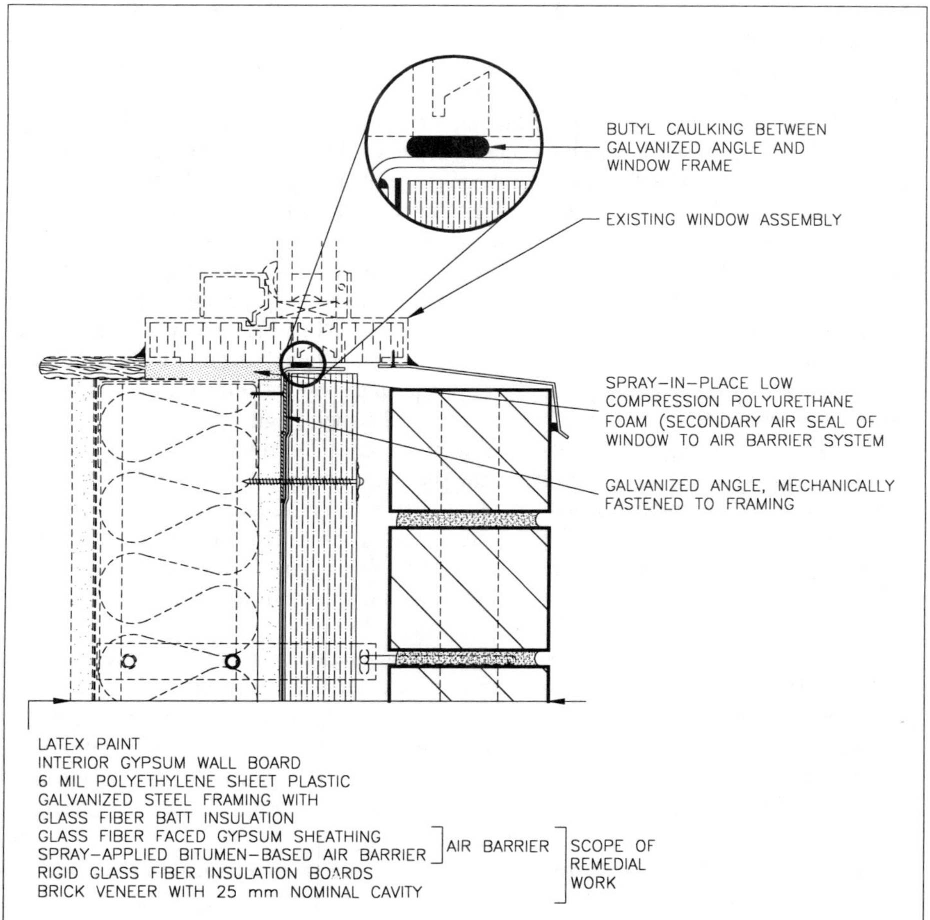

Figure 3 – *Window Sill Detail*

- Structural Support: it is essential that the proper structural support be provided to the air barrier connections, such that wind loads, deflection, and air pressure differentials do not cause the connections to fatigue and fail.

Further to the major component properties of the air barrier system, the connectivity of the system relies on two specific elements. In Figure 2, the connection of the window jamb assembly is illustrated in a flow chart. The arrows indicate the sequence of the material applied to the adjoining material.

A redundant connection of the window to the wall assembly, that accommodates construction tolerance, is accomplished first with sprayed-in-place polyurethane foam between the sheathing and the window frame (the core of which is insulated with rigid expanded polystyrene, inserted during manufacture). Secondly, a galvanized metal angle that is screw fastened to the steel framing, through the sheathing is caulked to the leg of the window frame with butyl, on the outer side of the thermal break. The metal angle is joined to the sheathing with self-adhering rubberized asphalt membrane, caulked on all terminations with butyl.

In Figure 3, the air barrier detail illustrates a similar interface between the air barrier and the window frame, however, as an added moisture management measure, the air seal is made between the window frame on the inner side of the thermal break, such that if any moisture does leak (although not necessarily anticipated), it can drip out through the thermal break (butt joint) intersection between the jamb and sill of the window frame.

In Figure 4, the main difference of this detail in comparison to the window sill and jamb details is the fact that a shelf angle supporting the brick veneer becomes a component of the air barrier system. The air seal is achieved by completely filling the void between the window head frame and the shelf angle with spray-in-place polyurethane foam. This also accommodates construction tolerance, i.e., proximity of the window frame to the shelf angle. In retrofit details such as this, it is not always possible to construct a fully redundant air seal, thereby requiring more keen inspection during the construction review. The air barrier is connected to the shelf angle above the window via the through-wall flashing detail (self-adhering membrane). Details were also required to provide continuity of the air barrier where the shelf angle terminated at the interface of the brick veneer with the curtain wall cladding, which also required end-dams to be formed into the through wall flashing component of the air barrier. The void between the curtain wall frame and air barrier were sealed with polyurethane foam, and the underside of the shelf angle was caulked with butyl to the adjoining curtain wall frame, similar to the window jamb detail illustrated in Figure 3.

The ability of spray-in-place polyurethane foam insulation to perform as effective air seal at window perimeters has been tested by Proskiw [*10*]. This material is easily applied, since it readily cures to fill to space into which it is applied, and adheres to most construction materials. The installation requires due care, and the material compatibility should always be reviewed with the foam manufacturers. In Proskiw's paper, the mean air leakage performance of five test specimens at 75 Pa air pressure was 0.0094 L/s•m, i.e., measured as L/s per linear meter of window perimeter.

The capability of low compression, low expansion polyurethane foam to resist air pressure differentials is typically acceptable. This material should demonstrate properties of tensile and compression strength in excess of 100 kPa, and should be capable of accommodating of some minor movement within the juncture, and not be capable of absorbing moisture beyond 5% of its relative mass. The properties of a polyurethane foam material in this type of detail, as a minimum, should conform with the Type 1 classification of ASTM Specification for Spray-Applied Rigid Cellular Polyurethane Thermal Insulation (C 1029).

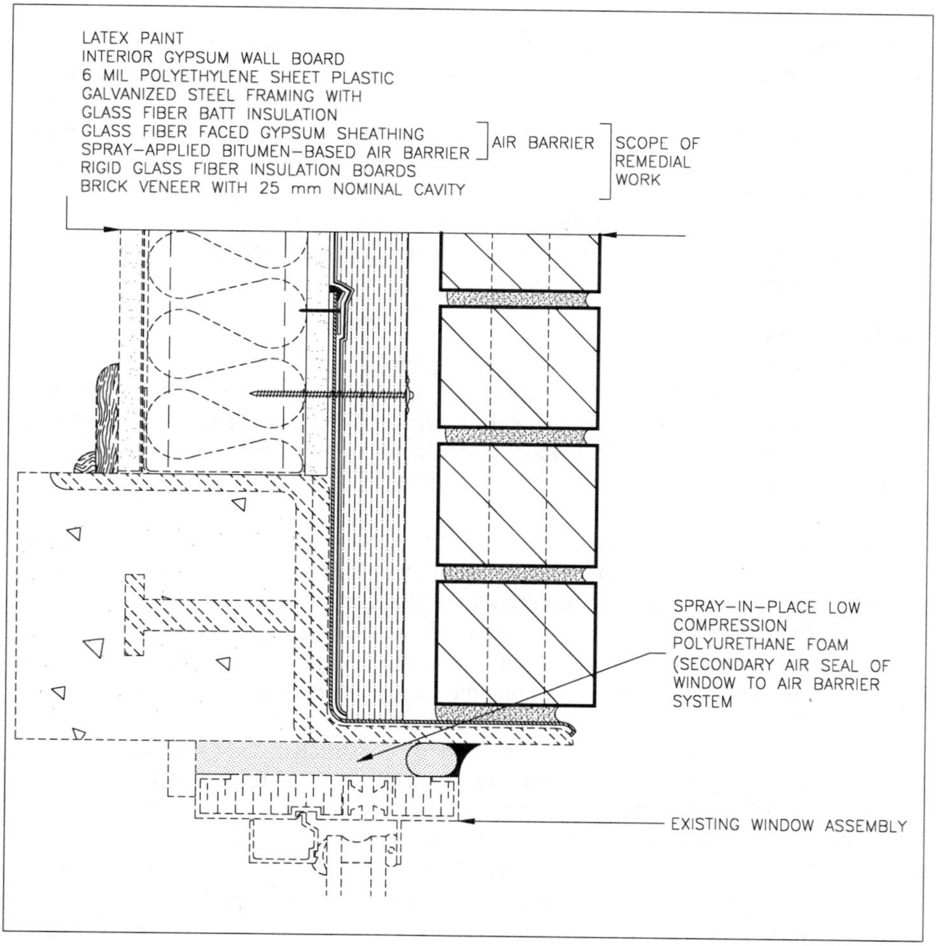

Figure 4 – *Window Head Detail*

In this example illustrated in Figures 1, 3, and 4, the wall sheathing is screwed to the steel framing, therefore, it is very important that the screw fastening not be such that the screw head penetrate beyond the plane of the glass fiber facing. The head of each screw should sit slightly proud of the surface of the sheathing, otherwise, screws set too far into the sheathing can drastically reduce the structural integrity of the sheathing since the gypsum core becomes displaced. Screws are required a maximum of 200 mm on centers, and if any screws penetrate the sheathing face, additional screws should be properly fastened on 100 mm to either side of the existing screw heads which may compromise the integrity of the sheathing boards.

The installation of the glass fiber faced sheathing also required that the board joints be filled with a polymer modified (moisture resistant) gypsum compound, and reinforced with glass fiber tape. The holes cut into the sheathing to accommodate the through-sheathing penetration of the bayonet-type brick ties was also filled with this compound. The continuity of the sheathing at the board joints and brick tie penetrations creates a well defined plane onto which the spray-applied bitumen mastic can be contiguously supported.

The self-adhering rubberized asphalt membrane component requires the use of primer for attachment to all surfaces, most especially for the glass fiber faced gypsum sheathing. All terminations and laps in the self-adhering membrane should be caulked with butyl to ensure that all seams, and any ripples (sometimes referred to as fishmouths) will be adequately sealed. The spray-applied mastic membrane overtop of these connections then ensures a complete air seal.

Quality Assurance

A methodology that may enable a designer to implement a comprehensive air barrier quality assurance program has been purported by the National Air Barrier Association (NABA) of Canada. This program has been established, based on the ISO9002 model for quality assurance, and it provides a basis upon which to qualify the variables for air barrier design and installation. In Figure 5, this model has been illustrated, based on NABA's published manual [*11*].

There is good potential for this NABA program to gain notoriety and be adapted into the construction marketplace, and it would certainly bolster the most critical education components required in constructing exterior walls, specifically:

1. Educate designers to provide accurate, and complete details for the building envelope of a given project, and recognize that the function of the air barrier is inter-connected with other principle requirements of the wall assembly.
2. Educate and certify contractors and inspectors to understand the importance of ensuring that critical details are properly constructed, and be capable of modifying details as required to address construction tolerances.

It could be debated that this program might make the process of designing and constructing a wall intrinsically more cumbersome. However, it should be recognized that the installation of air barrier systems can be sufficiently complicated, insofar as how the air barrier is installed, via different subtrades, and the common problems that arise from construction sequencing and scheduling issues. This model is industry based, hence, the aspects of "Database Tracking, Appeal Process, and Research & Development" are not specific to the design of a given project. However, the information collected from each project can eventually be used by industry to generate statistics, identify problem-areas, and provide certification and/or licensing of installers (individuals), contractors (corporate entities), manufacturers, and inspectors.

This program has great potential, and far-reaching implications towards addressing not only the air barrier system, but also all the principle requirements of a wall system.

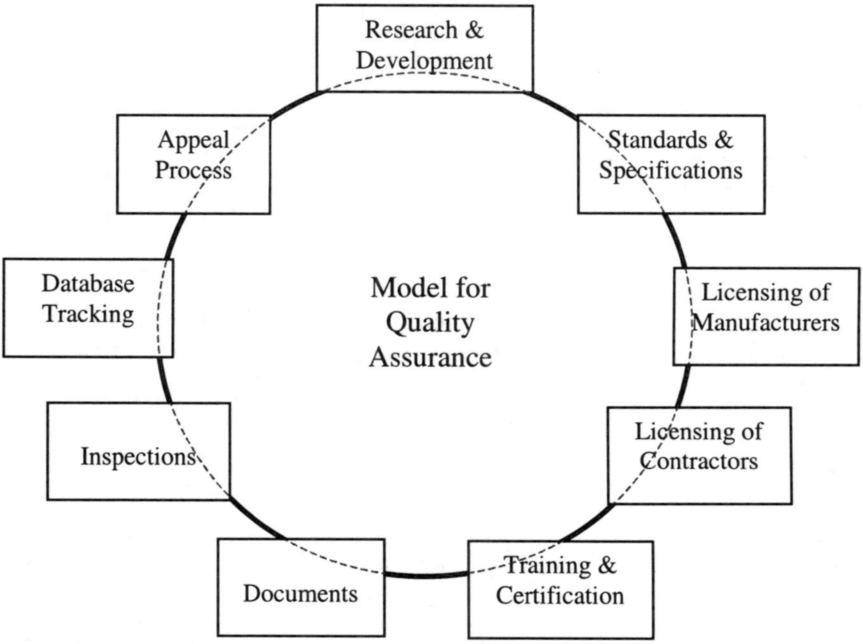

Figure 5 – *NABA Quality Assurance Model for Air Barrier Systems*

Construction Sequence and Scheduling

Often, it can be misperceived that adding some minor, simple procedures to the installation of the air barrier components will add tremendous costs and time to a given contract. However, the air leakage resistance of almost any wall assembly will be inversely proportional to the service life of the wall assembly or building system.

Therefore, the relatively minor additional costs (if any) for supplementary air seals in the air barrier system can be extremely cost effective.

Adopting the NABA model for quality assurance, insofar as the example wall assembly provided earlier, is illustrated in Figure 6. The correlation of the design, materials, and installation stems from the designer having the appropriate control of the construction process, insofar as contractor qualifications, contract administration, and construction inspection. In the scope of the construction of a new building, or the rehabilitation of an existing building, the owners may be all too eager to implement quick, low-cost solutions. The experience of many practitioners of building envelope consulting has revealed that most often, the simple details were never addressed during design, nor construction, thereby resulting in premature failure. The simplicity of constructing an air-tight wall assembly, as purported by this author, is contained within the necessity of detailing, and constructing air tight interfaces within the various junctures of a given wall assembly, ensuring that the materials are compatible and durable for the intended use.

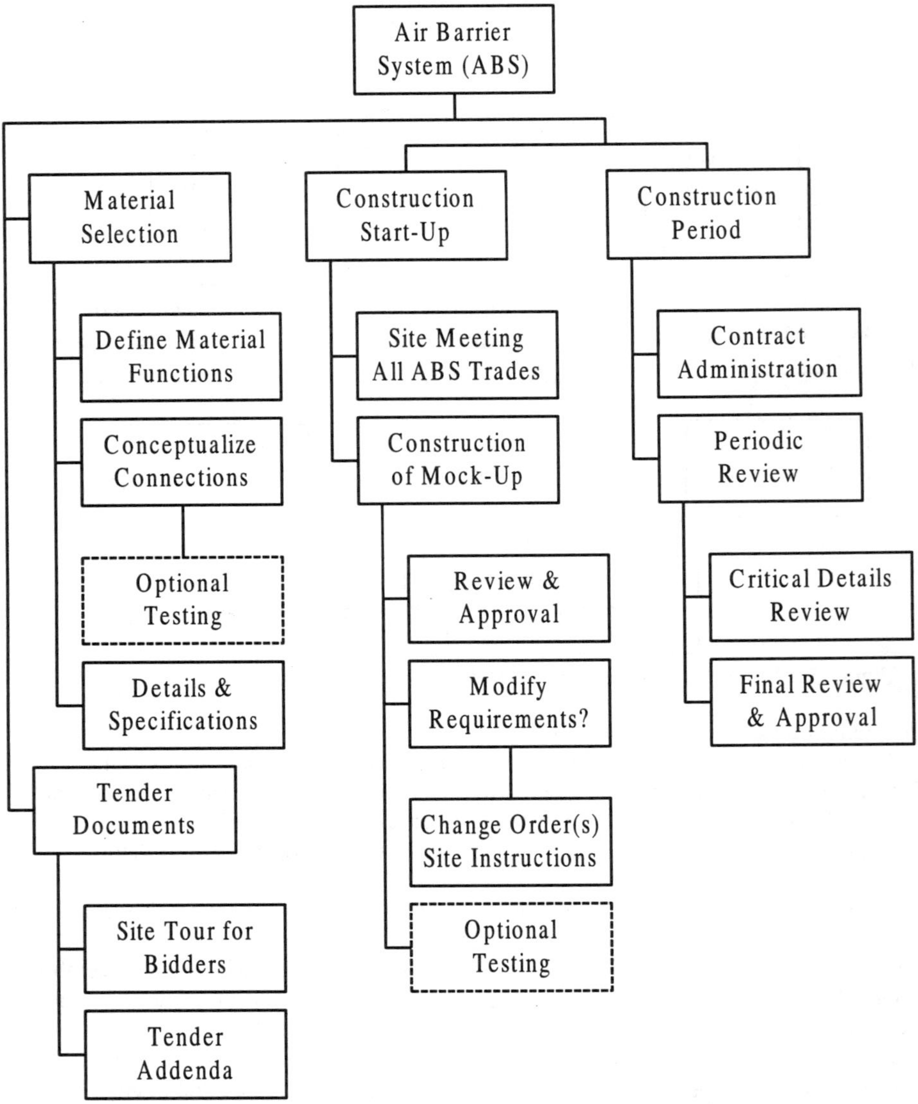

Figure 6 – *Simplified Quality Assurance Process for Air Barrier System*

Conclusions

In conclusion, the appreciation for understanding the intent of what an air barrier system is to achieve must give due consideration to the control of moisture, thermal transfer, and structural integrity.

The very definition of an air barrier system implies that it be continuous, hence, the connectivity of the materials which comprise the plane of air-tightness becomes the most critical aspect of the system. The selection of the appropriate materials, being compatible at the interfaces, must also account for designing these interfaces with buildability, and construction sequence in mind. The smallest, most critical details must be well-thought-out, and implemented accordingly. Further, even a foolproof design will require, at the very least, periodic inspection that should be conducted at the critical steps in the construction of the wall. Although not addressed herein this paper, there are methods and procedures for conducting field tests; however, these are limited to being representative of the given areas tested, and not necessarily representative of the entire air barrier that is built during construction. It is the opinion of this author that details which combine well-devised construction sequencing with durable material interfaces and redundant seals between critical material connections will ensure that the design intent for the air barrier can be properly implemented.

The concept of air barrier connectivity provides the basis with which the various interconnections can be assessed and designed to ensure that the air barrier system can perform its function satisfactorily. The essence of designing with connectivity in mind requires that the air barrier be designed as a series of materials and components, thereby forming a system.

References

[1] Hutcheon, N. B., "Requirements for Exterior Walls," Canadian Building Digest, Number 48, National Research Council of Canada, December 1963.

[2] Brown, W. C., Di Lenardo, B., Poirier, G. F., and Lawton, M. D., "An Evaluation Guide for Performance Assessment of Air Barrier Systems," ASHRAE Thermal Envelopes VII, American Society for Heating, Refrigeration and Air-conditioning Engineers, December 1998.

[3] Lstiburek, J. W., and Carmody, J., "Moisture Control Handbook," 1993, Van Nostrand Reinhold.

[4] Handegord, G. O. P., "Building Science and the Building Envelope" 1999, Handegord & Company Inc.

[5] Murray, S. M., "Solving Roof Leaks with Fans," Roofing Consultants Institute (RCI) Interface Magazine, October 2000.

[6] Straube, J. F. and Burnett, E. F. P., "A Review of Rain Control and Design Strategies," Journal of Thermal Envelope and Building Science, July 1999, pp. 41–56.

[7] Lawton, M. D., "Are We Sealing the Wrong Walls in Apartments?" Royal Architectural Institute of Canada Advanced Buildings Newsletter, Vol. 1, No. 4, July 1994.

[8] Day, K. C., "Exterior Insulation Finish Systems: Designing EIFS (Clad Walls) for a Predictable Service Life," 8th Canadian Conference on Building Science & Technology, February 2001.

[9] Brown, W. C., Di Lenardo, B., Poirier, G. F., and Lawton, M. D., "An Evaluation Guide for Performance Assessment of Air Barrier Systems," ASHRAE Thermal Envelopes VII, American Society for Heating, Refrigeration and Air-conditioning Engineers, December 1998.

[10] Proskiw, G., "Air Leakage Characteristics of Various Rough-Opening Sealing Methods for Windows and Doors," Airflow Performance of Building Envelopes, Components, and Systems, ASTM STP 1255, Mark Moderna and Andrew Persily, Eds., ASTM International, West Conshohocken, PA, 1995, p. 131.

[11] National Air Barrier Association Inc., "Professional Contractor Quality Assurance Program," 2nd ed., April 1997.

Ali M. Memari,[1] *Mohammad Aliaari,*[2] *and Ahmad A. Hamid*[3]

Evaluation of Seismic Performance of Anchored Brick Veneer Walls

Reference: Memari, A. M., Aliaari, M., and Hamid, A. A., **"Evaluation of Seismic Performance of Anchored Brick Veneer Walls,"** *Performance of Exterior Building Walls, ASTM STP 1422,* P. G. Johnson, Ed., ASTM International, West Conshohocken, PA, 2003.

Abstract: This paper reports the first part of an ongoing research project that is looking into the seismic performance of veneer walls. The type of veneer of interest for this work is normally anchored to the backup wall through metal ties. Brick veneer walls are supported in most cases by shelf angles attached to the floor slab at each story and are supposed to carry only their own weight and not participate in in-plane lateral load resistance. To achieve this behavior, horizontal and vertical movement joints are necessary. Ideally, this design could isolate the lateral movement of the backup wall from that of the veneer wall, thus preventing any distress to the veneer. However, earthquake reconnaissance reports show many failures of veneer walls with the potential of life-safety hazard. In this paper, it is discussed how the vertical differential movement between the brick veneer and the frame can close the gap between the underside of the shelf angle and the top course of brick, thus putting the brick veneer under high compressive stresses. It is shown that this can result in proportionally high friction forces during earthquakes with the possibility of shear cracking of the veneer before sliding between the brick veneer and the supporting steel shelf angle occurs.

Keywords: brick veneer, seismic performance, friction forces

[1]Assistant Professor, Department of Architectural Engineering, The Pennsylvania State University, University Park, PA 16802.

[2]Graduate Assistant, Department of Architectural Engineering, The Pennsylvania State University, University Park, PA 16802.

[3]Professor, Department of Civil and Architectural Engineering, Drexel University, Philadelphia, PA 19104.

Introduction

Brick veneer walls are usually used as the skin of exterior concrete masonry unit walls. The brick veneer is used as the exterior wythe of two component cavity walls with an air space between the veneer and the backup wall (concrete masonry unit or steel stud). The type of veneer of interest for this work is normally anchored to the backup wall through metal ties that transfer lateral (out-of-plane) wind and seismic loads to the usually stiffer concrete masonry unit backup wall. Brick veneer walls are supported in most cases by shelf angles attached to the floor slab at each story and are supposed to carry only their own weight and not participate in in-plane lateral load resistance. To achieve this behavior, horizontal and vertical movement joints are necessary. Ideally, the horizontal movement joints could allow the brick veneer in a given story to move in-plane relative to brick veneers in adjacent stories, thus preventing any distress to the veneer. However, earthquake reconnaissance reports show many failures of veneer walls with the potential of life-safety hazard. The problem can be traced to the transfer of in-plane vertical and seismic induced lateral forces from the backup wall (which deflects with the structural frame) and shelf angle to the veneer wall.

This paper reports the first part of an ongoing research project that is looking into the seismic performance of veneer walls with the objective of suggesting seismic isolation schemes. In this paper, the primary objective is to explore the mechanism of vertical and lateral in-plane force transfer from backup wall and floor supported shelf angle to brick veneer. It is discussed how vertical deformations due to elastic deflection of the frame, creep, temperature change, and moisture expansion of brick in walls can put into compression the brick veneer wall when the compressible filler or the open space at the horizontal joint between the underside of the shelf angle and the top course of brick is effectively closed. It is shown that this can result in potential transfer of vertical gravity loads from the shelf angle to the veneer wall. The paper discusses that such transfer of vertical loads can be accentuated during moderate to strong earthquakes due to lateral sidesway movement of the frame with the result of creating large friction forces between the shelf angle and the brick veneer. It is shown that the friction forces at closed horizontal control joints can increase the participation of the brick veneer wall in lateral load resisting and thus increase the potential of veneer failure.

Performance of Brick Veneer Walls in Past Earthquakes

Earthquake reconnaissance reports show failure of veneer walls with the potential of life-safety hazard. The Loma Prieta Earthquake reconnaissance report [1], which describes the damages incurred as a result of the October 17, 1989 earthquake near Santa Cruz, California, summarizes the observations of damage to brick veneer as follows: "Damage to exterior unreinforced masonry – brick veneer and façade systems, especially in upper stories, that resulted in extreme life-hazard to pedestrians below." The Northridge Earthquake reconnaissance report [2], which discusses the

Figure 1— *Masonry Veneer Damage at a Residential Building During Northridge Earthquake (from EERI [2]).*

effects of the January 17, 1994 earthquake near Los Angeles, California, gives a more detailed description of the types of damage and possible causes. According to this report, "anchored veneer experienced a large fraction of the damage observed to modern masonry" and that most of the damaged veneer cases had to do with insufficient anchoring system that ties the masonry veneer to the backup wall. Figure 1 shows an example of a brick veneer failure in Northridge Earthquake [2]. On the other hand, in cases where adequately sized movement joints had been constructed in the brick veneer, the performance has been acceptable. According to the Northridge Earthquake reconnaissance report, two buildings with partially constructed brick veneer over steel stud walls on the University of California at Los Angeles campus "had adequately sized and spaced movement joints in the brickwork, and both apparently performed well."

In a study of several buildings that sustained brick veneer wall damage during the Loma Prieta Earthquake, Jalil et al. [3] report that the damage ranged from diagonal cracks to some spalling to complete loss of the veneer wall. In the analytical studies that Jalil et al [*3*] performed on the selected buildings damaged in the earthquake, they found out that by including the stiffness of the veneer in their computer models, they could predict the observed failures. They thus concluded that brick veneer walls could have a significant effect on the seismic response of buildings.

Understanding the Behavior of Anchored Brick Veneer – A Literature Review

Brick veneer walls are generally of two types; adhesion veneer and anchored veneer. Adhesion veneers such as terra cotta or thin brick are so bonded to the back wall that they do not offer much resistance in the out-of plane direction but take part in resisting in-plane lateral forces. On the other hand, anchored veneers are usually separated from the back wall by an air space and are attached to the back wall with masonry ties such that the veneer takes part in out-of plane load resistance but is not supposed to take part in in-plane lateral force resistance when proper movement joints are used. According to Wintz and Yorkdale [4], when the height of the brick veneer is large or the number and location of openings dictates, it would be necessary for the brick veneer to be supported on shelf angles secured to the structural frame. In such cases, it is necessary that horizontal "pressure-relieving" joints be constructed beneath the shelf angle such that either an air space or a compressible material is provided there to permit the veneer to freely move in-plane relative to the frame. Figure 2 shows a typical detail [5] for a brick veneer wall with concrete masonry unit backup wall. It should be noted that the use of a two-piece anchor to tie the brick veneer to the backup wall is very common. Grimm [6] presents a review of various types of masonry ties along with their structural properties. The anchor ties should be flexible such that they resist out-of-plane tension and compression, but not in-plane shear forces. It is obvious that if movement joints are not used, as in Figure 3(a), or are not functional, as shown in Figure 3(b) [7], where the veneer can be in tight contact with the underside of the shelf angle, relative displacement between the veneer and the back wall will be prevented, leading to participation of the veneer in in-plane force resistance.

As an example for such a situation, Brock [8] mentions a case study, where as in Figure 3(b), instead of sealant, mortar was used in front of the shelf angle toe, which led to the spalling of the face of the brick in a few locations due to excess compression in the veneer. The project involved the replacement of the brick veneers of exterior walls and partially sloped roof of a building on a university campus in the Northwestern United States that cost over $6 million in 1994, while the cost of the entire building constructed in 1974 was just over $7 million, which is equivalent to approximately $18 million in 1994. The construction, which according to Brock "was typical for the early 1970's," did not have horizontal expansion joints below shelf angles, neither did it have any vertical expansion joints near corners. The accumulated vertical compressive stresses in the brick veneer wall can lead to spalling of the face of the brick under the shelf angle, as shown in Figure 4 [7]. The vertical pressure buildup is the result of several sources, which are discussed next.

Hamid et al. [9] present a comprehensive review of the sources of vertical deformation in concrete frames and brick veneers and the differences in such deformations that lead to excessive vertical compressive stresses on the veneer. Accordingly, axial shortening of concrete frames are due to elastic deformation of columns and spandrels under load and the deformation due to shrinkage and creep. According to Hamid et al. [9], the overall frame shortening, including the three mentioned effects, can range from 0.01% to 0.09%. However, clay brick veneer

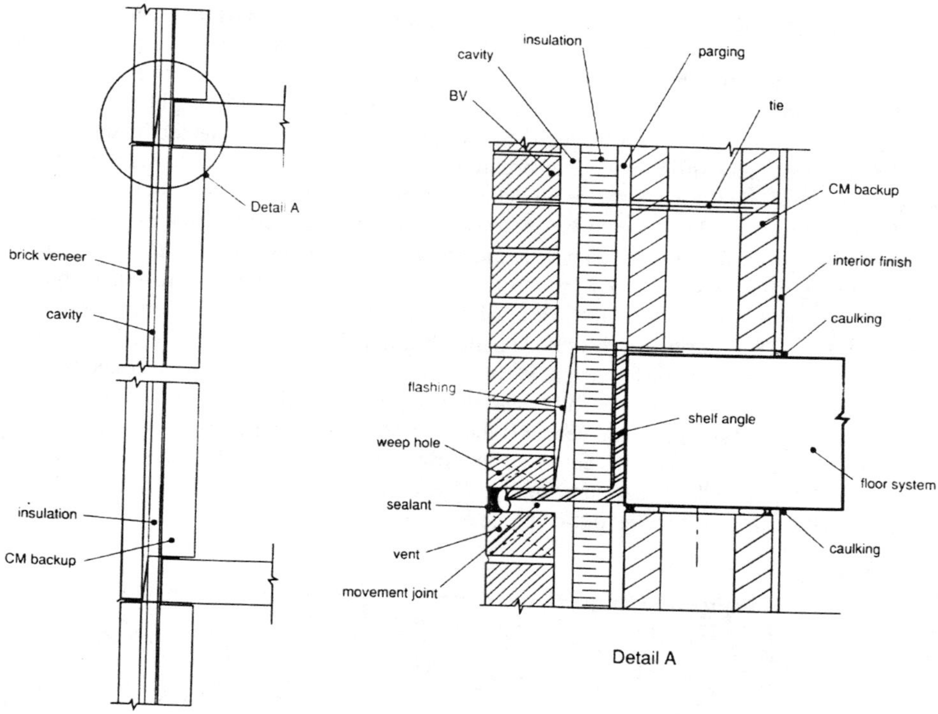

Figure 2 — *Typical Detail of Brick Veneer – Concrete Masonry Wall System (from Drysdale and Suter [5]).*

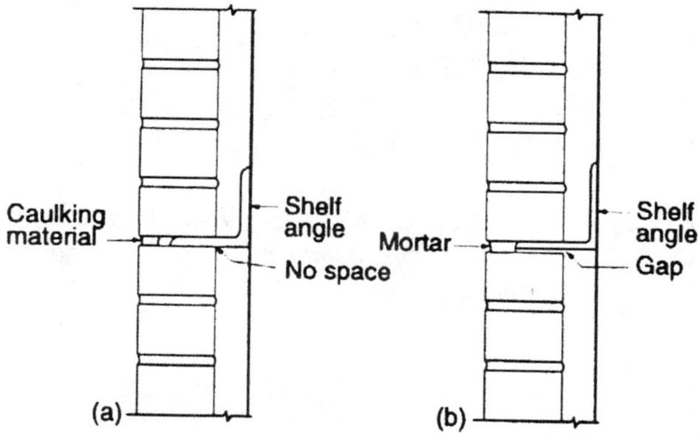

Figure 3 — *Veneer on Shelf Angle with No Movement Joint (from Drysdale et al [7]).*

walls have negligible elastic deformation because of the small loads they (are supposed to) carry, i.e., veneers are usually designed to carry their own weight, and also have negligible shrinkage compared to cementitious material. Moreover, creep deformations are negligible for the same reason as the elastic shortening because of the low axial stress level. However, the clay brick is significantly affected by the temperature change from that at construction time to the peak temperature at post construction, and for cases where no detail information is available, a horizontal thermal expansion 0.045% may be assumed [9]. Typical average thermal expansion coefficient for brick masonry can be assumed to be in the range of 2.5×10^{-6} to 4.0×10^{-6} in./in./ °F (0.0045 to 0.0072 mm/mm/ °C) [7]. Finally, brick moisture expansion, with a range of 0.016% to 0.028% according to Ritchie [10] and with a recommended value of 0.02% according to Monk [11] and Grimm [12], can also significantly affect brick movement. The difference in vertical deformation between the veneer and the structural frame system due to the described sources can lead to significant vertical forces being transferred to the veneer if the horizontal movement joint is not properly designed and functional. This can result in bowing or buckling of the brick veneer as shown in Figure 5. Such a case is highly vulnerable in an earthquake due to a combination of in-plane and out-of-plane earthquake forces.

Figure 4 — *Spalling of Brick Veneer at Shelf Angle Under High Compressive Stresses Due to Lack of Masonry Joint Under the Shelf Angle (from Drysdale et al [7]).*

Hamid et al. [9] presented the results of a study on a 6-story apartment building consisting of reinforced concrete flat plate and column construction that had bowing of the masonry veneer and some spalling at the shelf angle locations. In this veneer wall the shelf angle, which was bolted rigidly to the spandrel slab, was under bearing pressure from the brick on the topside and the brick at the underside of the "relieving angle" because of the closure of the horizontal joint. Hamid et al [9] showed that the concrete frame had a total deformation (shortening) of 13.6 mm due to elastic shortening, creep and shrinkage in the frame behind one panel for one story (2.6 m high). At the same time, according to Hamid et al. [9], the net deformation (extension) in the brickwork panel resulting from the creep (shortening), thermal movement and moisture growth was 3.2 mm for the entire height of the brick panel (2.6 m). The differential movement between the frame and the brick veneer produced a compressive strain of 0.0012, which was equivalent to a stress of 14.1 N/mm^2 (MPa). According to Hamid et al [9], this stress was large enough (58% of the ultimate compressive strength of masonry) to cause spalling of the veneer.

Figure 5— *Buckling of Brick Veneer Wall Due to Vertical Differential Movement between Veneer and Frame (from Drysdale et al [7]).*

Statement of the Problem

Given the insight into the behavior of brick veneers through the literature review, the objective of this study is to investigate the potential for failure of brick veneer walls for cases such as those demonstrated in Hamid et al.'s work. In other words, the question is how would anchored brick veneer walls with ineffective horizontal pressure-relieving joints, that is, closed under high compressive forces or simply nonexistent, behave in an earthquake. Hamid et al.'s work indicates that in some cases, the veneer wall could be subjected to large vertical compressive stresses that can lead to bowing of the wall and spalling of the masonry under the static effects of elastic shortening, creep, shrinkage, temperature and humidity. If under such circumstances, the building also experiences an earthquake, what could happen to the brick veneer wall? This is the question that is addressed in this study.

The procedure to investigate this issue consists of determining the friction force that will be developed between the shelf angle and the brick veneer when the building tends to displace laterally (i.e., parallel to the brick veneer wall). It is assumed that the brick veneer is subjected to large vertical compressive stresses in the range determined by Hamid et al. [9]. As the building frame tends to displace laterally in an earthquake event, its movement will be resisted by the friction resistance between the shelf angle and the brick veneer. This resistance forces the brick veneer wall to take part in resisting lateral seismic loads until either slip occurs between the shelf angle and the brick or until some failure mechanism occurs in the brick wall. The friction coefficients are obtained using the test results reported by McGinley and Borchelt [13]. The calculations for the work reported in this study were performed by hand.

Description of the Model for Analysis

The model considered for structural calculations is a 20 ft (6096 mm) long by 12 ft (3658 mm) high single-wythe brick veneer wall (90 mm thick) anchored to a concrete masonry backup wall. The brick veneer wall is assumed to be in contact with the shelf angles at top and bottom and is under compressive stresses resulting from differential vertical deformation of the frame and the brick veneer wall. Since the shelf angle is assumed to be anchored into the floor slab, the lateral building displacement is restrained by the friction forces at the brick-shelf angle interfaces. This results in a coupling of the structural frame and the veneer wall in resisting inter-story lateral displacements.

Friction Force

Because the friction force plays a significant role in the in-plane seismic response of brick veneer walls under compression force, relevant work reported by McGinley and Borchelt [*13*] is next reviewed. Prior to this work, which was commissioned by the Brick Institute of America, the coefficient of friction that was used for masonry was apparently based on the values for reinforced concrete. In that sense, the contribution of McGinley and Borchelt in determining friction coefficient at the interface of brick and concrete, with and without flashing, and at the interface of brick and shelf angle with and without flashing is significant. McGinley and Borchelt tested several different cases of shelf angle support under the brick veneer that are of primary interest in the present study. The specimens for the 31 tests that were supported on shelf angles were wallettes consisting of three units long and either three or four courses high. An axial load of 600 lb (2670 N) was exerted on each wallette while either in-plane or out-of-plane force was applied to cause slip to occur. Two commonly used types of flashing material were used in these tests, 30 mil (0.76 mm) PVC flashing and 3 oz/ft^2 (915 g/m^2) paper-backed copper flashing (paper on the support side). For each configuration five tests were performed in order to provide reliable average values for friction coefficients. This study resulted in an average value for the coefficient of friction between 0.6 to 0.7 for various cases with shelf angle supporting brick veneer. The coefficient of friction was slightly reduced with increased axial force to a lower limit of approximately 0.6. The authors also cite values of 0.3 to 0.4 for coefficient of friction for masonry supported on metals as reported by Amrhein [14]. While the latter values are more conservative, the recent test results by McGinley and Borchelt seem to be more reliable for brick veneer with the use of PVC or paper-backed copper flashing between brick and steel shelf angle. Figure 6 shows the variation of friction coefficient with axial load for both cases of steel shelf angle support and concrete support.

While most of the tests were performed by pulling the wallettes longitudinally under an axial load of nearly 600 lb (2670 N), two sets of tests (five per set) were also carried out under axial loads of approximately 100 lb (445 N) and 1500 lb (6672 N) to show the effect of the change of axial load on friction coefficient. Moreover, to see the effect of in-plane force versus out-of-plane force on the friction coefficient, one set of test was also carried out under the axial load of 600 lb (2670 N) by pulling the wallettes transversely. The average of each set of (five) test for various configurations is shown as a dot in Figure 6. For a meaningful structural analysis of a brick veneer wall system, proper boundary conditions should be taken into account, and under large axial forces developed in the veneer wall, the friction force should be considered. It should also be mentioned that other parameters such as the condition of contact surfaces and flashing and the existence of other elements (e.g., windows) in the assembly can influence the analysis results, and can be considered in the analysis for further refinement.

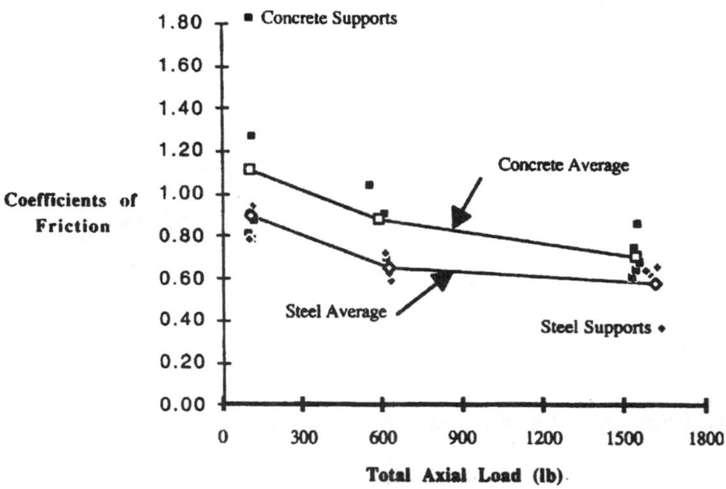

Figure 6 — *Variation of the Friction Coefficient with Axial Load (from McGinley and Borchelt [13]).*

Considering the 20 ft (6096 mm) by 12 ft (3658 mm) model of a brick veneer wall with a thickness of 90 mm and average values of coefficient of friction from Figure 6, friction forces for various values of axial loads on the veneer wall can be determined. The range of axial loads on the brick veneer wall model can be obtained from the results of the study by Hamid et al. [9] or the studies that will be mentioned subsequently. The variation of the friction force with axial force for each average value of friction coefficient is plotted in Figure 7 for steel supported brick veneer. It should be noted that as shown in Figure 2, the brick veneer is projected outward from the shelf angle a short distance recommended [7] to be less than 1 ¼ in. (30 mm) or 1/3 the brick veneer thickness. That means the actual contact area between the brick and shelf angle is less than the full area of the base of the brick veneer wythe. This refinement, however, was not considered in the calculation shown in Figure 7 and full area has been used. To evaluate the implications of such large values of friction forces, we need to estimate the lateral load that initiates cracking in a masonry wall. Typical failure loads for masonry walls can be estimated by reviewing the available test results reported in the published literature.

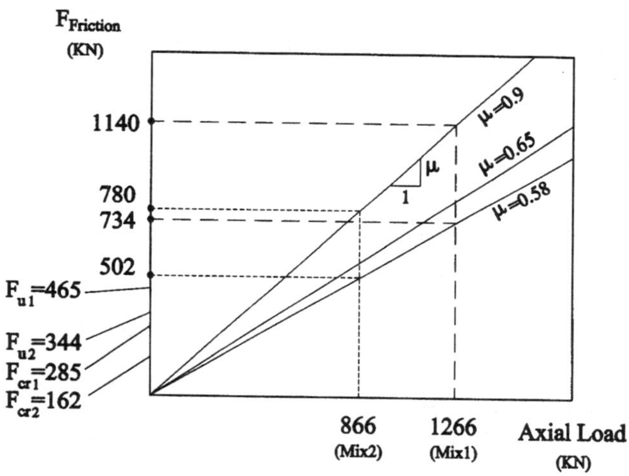

Figure 7 —*Friction Force-Axial Load Relation for the Model Brick Veneer Wall.*

Lateral Load Capacity of Unreinforced Masonry Walls

The problem under investigation manifests itself in some of the existing and older buildings. Therefore, any test results that indicate the strength of older unreinforced masonry walls will be very useful for this study. In an interesting study, Abrams [15] reports on testing the masonry walls of a building built in 1917. Five walls of this building were transported to the laboratory for lateral load testing. Vertical compressive stresses in the range of 0.52 to 0.99 MPa (76 to 143 psi) were applied to the walls. Figure 8 shows the results of the lateral tests performed by Abrams [15]. The shear stress plotted was obtained by dividing the lateral force by the gross wall area. The flexural cracks were observed at 40% of the ultimate load, which according to Figure 8 occurred at shear stress in the range of 0.41 MPa (60 psi) to 0.55 MPa (80 psi).

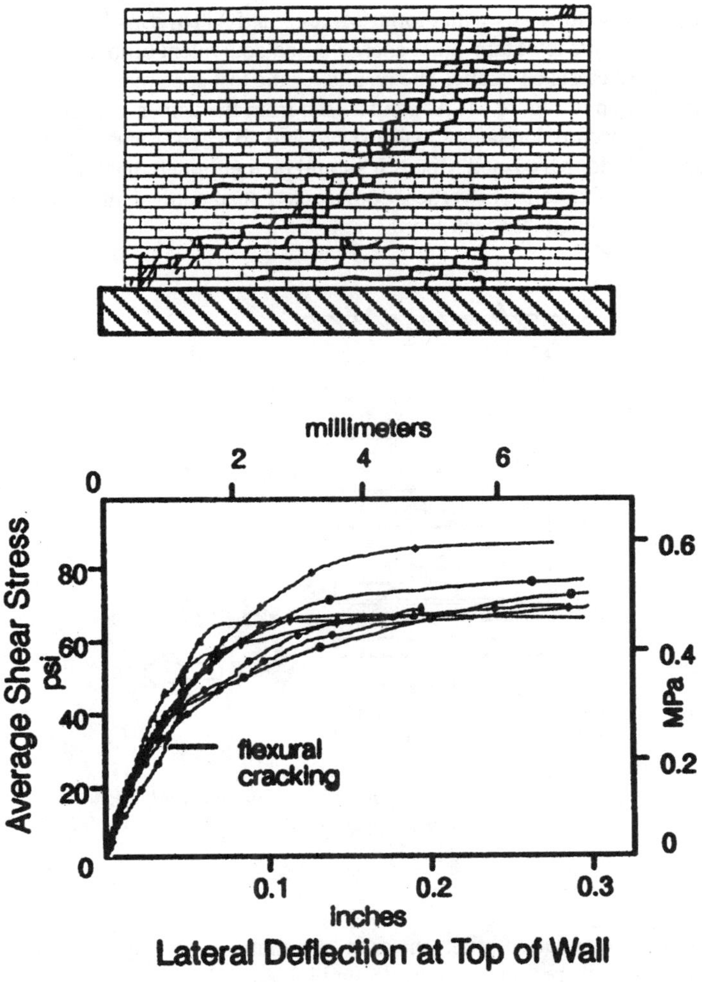

Figure 8 — *Lateral Load Deflection Test Results on Brick Veneer Walls (from Abrams [15]).*

The cracking load at 40% of the ultimate load obtained by Abrams [15] is not much different from the results obtained by Bosiljkov et al. [16] in their experimental study of a series of unreinforced masonry walls. The objective of Bosiljkov et al's research was to study the effect of different mortars on the shear strength of masonry walls. From among the various mortar mixes that the authors used, Mix 1 containing cement: sand in volume proportion of 1:4 and Mix 2 with cement: lime:sand in volume proportion of 1:1:6 are of interest for this study. Figure 9 shows a typical failure mode of test panels [16], and Figure 10 shows hysteresis loops of lateral force – displacement response for walls with mortar Mixes 1 and 2 [16]. Bosiljkov et al. then used these test results to obtain equal energy-based equivalent bilinear idealization of the hysteresis envelope. The results for the test specimens with two mortar mixes are shown in Figure 11. As can be seen from the values in Figure 11, the flexural cracking occurred at lateral loads 61.6% and 46.8% of their maximum strengths, respectively for Mix 1 and Mix 2. It should be mentioned that the lateral load tests were done under constant vertical compressive load equivalent to 1/6 of the wallette compressive strength for each mortar mix.

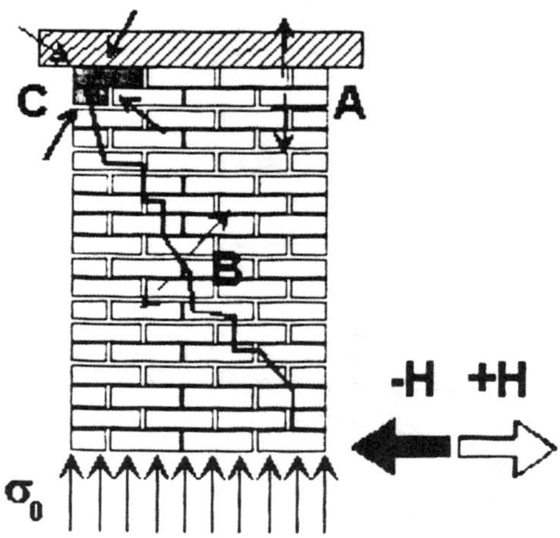

Figure 9 — *Failure Mode for a Brick Shear Panel (from Bosiljkov et al. [16]).*

Discussion

The objective in comparison of brick veneer in-plane lateral load capacity with the friction force, for cases where there is no horizontal joint and the veneer is under compressive stresses, is to predict the mode of failure in an earthquake. If we use the compressive stresses that cause spalling of the brick veneer as reported by Hamid et al. [9], that is 14.1 MPa (2045 psi), the compressive force on the shelf angle (assuming full contact area) for a horizontal wall section of 6096 mm by 90 mm will be 7736 kN (1739 kips), which results in a potential friction force of 4487 kN (1009 kips) for the

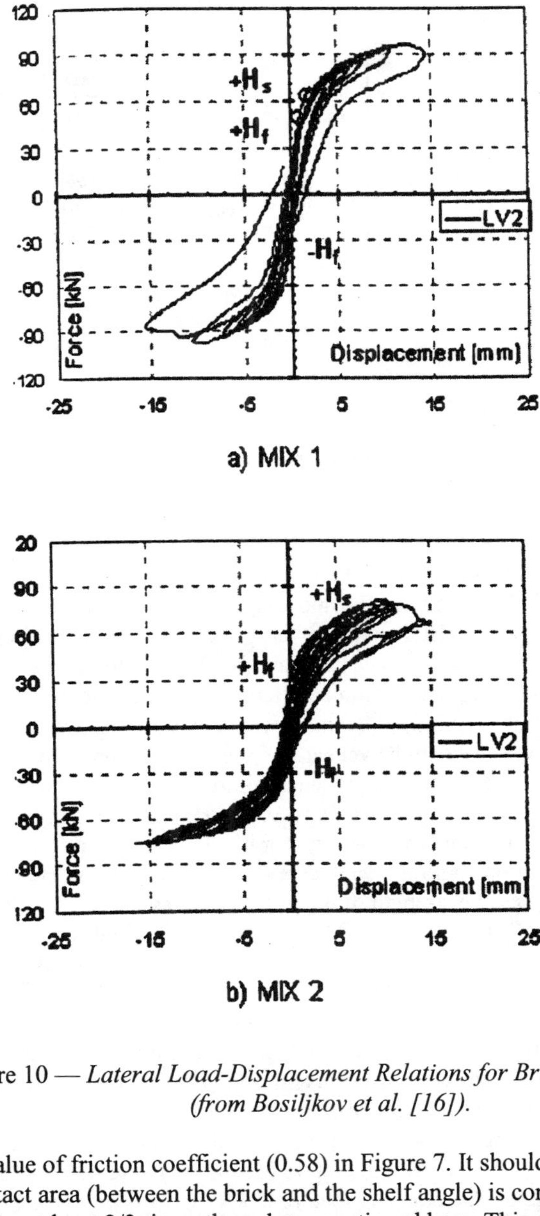

Figure 10 — *Lateral Load-Displacement Relations for Brick Shear Panels (from Bosiljkov et al. [16]).*

smallest value of friction coefficient (0.58) in Figure 7. It should be noted that if the actual contact area (between the brick and the shelf angle) is considered, the friction force will be at least 2/3 times the values mentioned here. This value of friction force is an order of magnitude larger than the cracking and ultimate capacity of the brick veneer wall. This indicates that a brick veneer under such level of compressive stresses will easily fail due to in-plane shear in an earthquake. In other words, shear

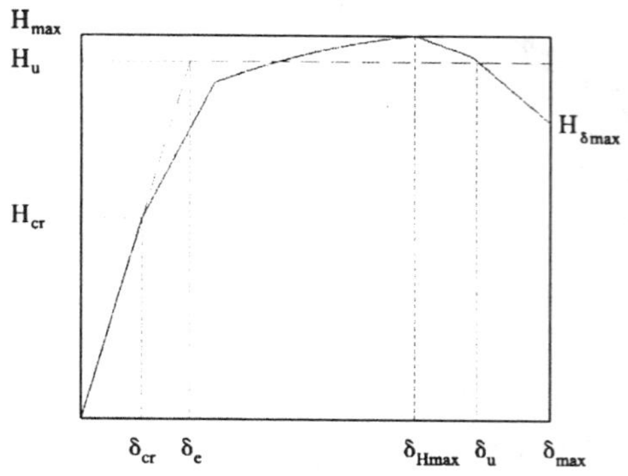

		MIX 1	MIX 2
H_{cr}	(KN)	59.5	33.5
δ_{cr}	(mm)	1.5	0.84
H_{max}	(KN)	96.6	74.6
δ_{Hmax}	(mm)	8.3	10.4
$H_{\delta max}$	(KN)	65.6	67.5
δ_{max}	(mm)	11.7	12.0
H_u	(KN)	90.0	63.8

Figure 11 — *Bilinear Idealization of Hysteresis Loops Based on Bosiljkov et al.'s Tests (Data from Bosiljkov et al [16]).*

capacity of the brick veneer will be reached before any sliding at the shelf angle takes place. Of course, it should be added that the anchorage of brick veneer to the back up wall will provide out-of-plane resistance. However, given that masonry ties are not generally designed to resist in-plane forces (they are flexible parallel to the wall direction), their contribution in the direction parallel to the brick veneer will be minimal in resisting seismically induced lateral forces. This analysis points out the potential for in-plane shear failure in the veneer. Existence of masonry ties may prevent fallout of the brick veneer in case of shear failure.

Next, we can assume smaller compressive stresses in the brick veneer. In order to have a sound basis for comparison, the brick veneer model in this study can be assumed to be under the same compressive stresses as the specimens in Bosiljkov et al tests, i.e., 1/6 the compressive strength of the prism, which were 13.85 MPa (2009 psi) for Mix 1 and 9.47 MPa (1373 psi) for Mix 2. The axial load information provided in Bosiljkov et al. [16] was translated into equivalent compressive force for the brick veneer model in this study. The values corresponding to Mixes 1 and 2 are 866 kN (195 kips) and 1266 kN (285 kips), respectively. By drawing vertical lines at these two points on Figure 7, friction forces corresponding to a range of friction coefficients can be read. For instance, if we take the smallest friction coefficient of 0.58, the corresponding friction force for Mix 2 equivalent is 502 kN (113 kips). This value is 3.1 times the cracking load (162 kN) and 1.5 times the ultimate load (344 kN). Similarly, the friction force corresponding to Mix 1 axial load level is 734 kN (165 kips), which is 2.6 times the cracking load (285 kN) and 1.6 times the ultimate load (465 kN). This analysis then shows that it is quite possible for the cracking or even ultimate lateral strength to be reached before any sliding takes place.

Still another comparison can be made using the data provided by Abrams [15]. The compressive stresses on the wall during the tests varied from 0.52 to 0.99 MPa, or an average of 0.75 Mpa (109 psi). For the brick veneer wall under study, this results in a compressive force of 412 kN (93 kips). If we assume a friction coefficient of 0.58, the potential friction force under this normal force will be 240 kN (54 kips). According to the data provided in Figure 8, cracking load (approximately 40% of the ultimate load) occurs at an average stress of 33 psi (0.227 MPa), which gives a lateral force of 125 kN (37 kips) for the wall model in this study. The ultimate lateral capacity of the wall will then be approximately 313 kN (70 kips). According to this analysis, the potential friction force is 1.9 times the cracking load but 77% of the ultimate load. This again verifies the previous result that in an earthquake, cracking of the veneer can occur before sliding if the veneer wall is under vertical compressive stresses large enough to cause spalling or bowing of the veneer.

Conclusion

An analysis of available data in several publications points to a potential problem posed by brick veneer walls with lack of horizontal movement joints when exposed to seismic loading. It should be pointed out that although some of the data used are based on experiments on wallettes, the use of resulting friction coefficients and shear stresses for full size walls is valid since such tests are carried out with the objective of using the results for real structures. To obtain statistically appropriate results, tests must be repeated several times, and use of wallettes is an economically feasible solution. There are many buildings with brick veneer walls with deficient horizontal joints, which result in high compressive stresses in the veneer due to differential movement. Based on available test results and structural calculations, this paper has indicated that the friction force under such circumstances can potentially exceed the cracking capacity of the wall and can be comparable to the ultimate shear strength of the brick veneer wall. The friction force can prevent sliding of the brick with respect to the shelf angle and in that case will let the brick veneer participate in lateral load resistance in an earthquake. This function (participation in in-plane lateral load resistance) has obviously not been considered in the design of existing brick veneer walls, and therefore the potential for failure of such walls in earthquakes cannot be ignored. For verification of the findings of this paper, a full-scale experimental study is recommended for brick veneer walls typical of older US practice. The recommendation for such cases (before any experimental verification) is that in moderate to high seismic regions, some preventive measures be taken for brick veneer walls that show any signs of distress due to differential movement of the brick veneer and the structural frame. It should also be added that if horizontal joints exist but with inadequate thickness, the level of compressive stresses in those joints can be expected to be smaller than the cases with brick veneer "locked in" at top and bottom as discussed in this paper. In these cases, therefore, the friction force that can develop at the interface between brick veneer and shelf angle will not be as high as discussed before. The current level of information and guidelines available for design and construction of brick veneer walls, including horizontal joints, is generally adequate. The concern raised in this paper is valid for some older buildings. If the problem is recognized in an existing building, masonry consultants can offer a variety of remedial solutions.

Acknowledgment

Major funding for this study was provided by the National Science Foundation under Grant No. CMS-9983896 as part of the first author's NSF CAREER award. The support of NSF is gratefully acknowledged. The opinions, findings, and conclusions expressed in this paper are those of the writers and do not necessarily reflect the views of the NSF.

References

[1] Earthquake Engineering Research Institute (EERI), "Loma Prieta Earthquake Reconnaissance Report," *Earthquake Spectra*, Supplement to Volume 6, May 1990.

[2] Earthquake Engineering Research Institute (EERI), "Nothridge Earthquake Reconnaissance Report, Vol. 2," *Earthquake Spectra*, Supplement C to Volume 11, January 1996.

[3] Jalil, I., Kelm, W., and Klingner, R. E., "Performance of Masonry and Mansonry Veneer Buildings in the 1989 Loma Prieta Earthquake," *Proceedings, The Sixth North American Masonry Conference*, Philadelphia, PA, 6–9 June, 1993, pp. 681–692.

[4] Wintz, J. A., III, and Yorkdale, A. H., "Brick Veneer Panel and Curtain Wall Systems – A Designer's Guide," *The Construction Specifier*, December 1983, pp. 24–35.

[5] Drysdale, R. G. and Suter, G. T., "Exterior Wall Construction in High-Rise Buildings – Brick Veneer on Concrete masonry or Steel Stud Wall Systems," *Canada Mortgage and Housing Corporation*, Canada, 1991.

[6] Grimm, C. T., "Masonry Veneer Anchors and Cavity Wall Ties," *The Masonry Society Journal*, August 1993, pp. 6–16.

[7] Drysdale, R. G., Hamid, A. A., and Baker, L. R., "Masonry Structures – Behavior and Design," 2nd Ed., *The Masonry Society*, Boulder, CO, 1999.

[8] Brock, L., "Design for Durability: Case Study of Anchored Brick Veneer that Meets Technical and Aesthetic Requirements," *Proceedings, Seventh North American masonry Conference*, South Bend, IN, 2–5 June, 1996, pp. 117–128.

[9] Hamid, A. A., Becica, I. J., and Harris, H. G., "Performance of Brick Veneer Masonry," *Proceedings, Seventh International Brick masonry Conference*, Melbourne, Australia, February 17–20, 1985, pp. 321–331.

[10] Ritchie, T., "Moisture Expansion of Clay Bricks and Brickwork," *Division of Building Research Publication No. 103*, NRCC, Ottawa, Ontario, 1975.

[11] Monk, C. B., "Analysis of Nonstructural Volume Changes in Masonry Construction," *Proceedings of the Fifth International Brick Masonry Conference*, Washington, D.C., October 1979.

[12] Grimm, C. T., "Designing Brick Masonry Walls to Avoid Structural Problems," *Architectural Record*, Vol. 162, No. 5, October 1977, pp. 125–128.

[13] McGinley, W. M. and Borchelt, J. G., "Friction at Supports of Clay Brick Walls," *The Masonry Society Journal*, February 1991, pp. 73–81.

[14] Amrhein, J. E., "Reinforced Masonry Engineering Handbook," 3rd Ed., *Masonry Institute of America*, Los Angeles, CA, 1987.

[15] Abrams, D. P., "Masonry as a Structural Material," *Proceedings of the Materials Engineering Congress, Materials Engineering Division, ASCE*, Atlanta, GA, 10–12 August, 1992, pp. 116–129.

[16] Bosiljkov, V., Zarnic, R., and Bosiljkov, V. B., "Shear Tests of the URM Panels Made from Different Types of Mortar – An Experimental Study," Proceedings, *12th International Brick/Block Masonry Conference*, Madrid, Spain, 25–28 June, 2000, pp. 303–317.

Andreas T. Wolf[1] and Pierre Descamps[1]

Determination of Poisson's Ratio of Silicone Sealants from Ultrasonic and Tensile Measurements

---

Reference: Wolf, A. T. and Descamps, P. **"Determination of Poisson's Ratio of Silicone Sealants from Ultrasonic and Tensile Measurements,"** *Performance of Exterior Building Walls, ASTM STP 1422*, P. G. Johnson, Ed., ASTM International, West Conshohocken, PA, 2003.

Abstract: Poisson's ratio is a fundamental material constant arising from the equations of linear elasticity and is a primary input property for finite-element analyses. Elastomers are generally regarded as nearly incompressible materials with Poisson's ratios slightly below 0.5. However, incorporation of particulate fillers may reduce Poisson's ratio significantly. Determining the Poisson's ratio of silicone sealants is an essential task, since these materials are increasingly used as structural adhesives in construction applications, which frequently require finite-element design calculations. The paper reports on the results of a feasibility study aimed at determining Poisson's ratio of three filled RTV silicone sealants based on simple ultrasonic and tensile measurements. Using this method, longitudinal ultrasound velocities of 984-1003 m/s were determined; and from Young's moduli and specific densities, transversal ultrasound velocities of 64-90 m/s and Poisson's ratios between 0.496 and 0.498 were calculated for the three sealants.

Keywords: Poisson's ratio, Young's modulus, ultrasound velocity, silicone sealant

Nomenclature

ν	Poisson's ratio
E	Elastic modulus
G	Shear modulus
K	Bulk modulus
L	Effective modulus for longitudinal waves
ρ_s	Specific density of material
v_L	Longitudinal ultrasonic wave velocity
v_T	Transversal ultrasonic wave velocity

---

[1] Scientist and senior process engineering specialist, respectively, Construction Industry, Dow Corning S.A., Parc Industriel, 7180 Seneffe, Belgium.

Introduction

Poisson's ratio is a fundamental material constant arising from the equations of linear elasticity and is a primary input property for finite-element analyses. Its origin comes from the remarkable property of classical solids to contract laterally in response to an applied normal (axial) stress. For infinitesimal strains, Poisson's ratio, ν, is the ratio between lateral (transverse) strain and axial strain during axial loading (Figure 1).

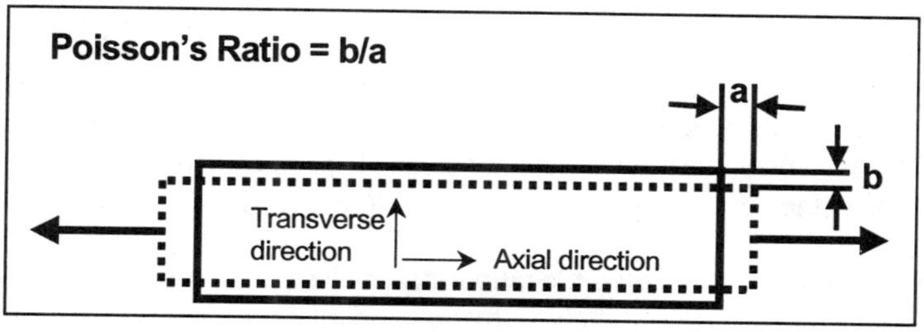

Figure 1 – *Definition of Poisson's Ratio*

For incompressible materials, this value is 0.5. Elastomers are generally regarded as nearly incompressible materials with values of ν slightly below 0.5. Incorporation of particulate fillers reduces Poisson's ratio, as can be numerically estimated from classical theories of rigid spheres embedded in an elastic matrix [*1,2*]. For instance, Holownia [*3*] measured $\nu = 0.49986 \pm 0.0001$ for unfilled natural rubber and noted the expected decrease with the addition of carbon-black filler. Waterman [*4*] studied sodium chloride filled polyurethane elastomers and found a decrease of about 0.05-0.1 in Poisson's ratio over the studied temperature range (-75°C to -30°C) by increasing the filler content from 0% to about 50% by volume.

Although the difference between these values and 0.5 may at first glance appear trivial, they are critical for many design calculations. This is because equations giving the stresses in a body frequently include the term $\nu/(1-2\nu)$ [*5*]. Since values of ν for elastomers are close to 0.5, small errors in ν can induce large errors in this term and, therefore, in the predicted stress values. To understand the significance of error propagation in structural design calculations, the finite-element work of Gent and Hwang [*6*] may be considered. These authors allowed ν to vary between 0.45 and 0.4999 during a calculation of a simple elastomer problem. Their results showed that the calculated stresses were extremely sensitive to the selected value of Poisson's ratio. In one sample problem, a 40% difference in calculated stiffnesses was found for $\nu = 0.49$ versus $\nu =$ 0.4999. It should be noted that their analyses ignored both time-dependent behavior as well as non-linear effects, which could be expected to produce even greater deviations.

Precise laboratory measurements of Poisson's ratio have proven to be difficult to obtain. Many of the classical experimental techniques are laborious and fraught with statistical uncertainties. This is because Poisson's ratio is a property, which for elastomers is not normally determined directly from experiment, but is deduced from any two of the following properties: elastic modulus, *E*, shear modulus, *G*, and bulk modulus,

K. For perfectly elastic, homogeneous and isotropic materials, these three properties are inter-related by equations (1a) to (1c).

$$\nu = E/(2G) - 1 \quad (1a)$$

$$\nu = \tfrac{1}{2} - E/(6K) \quad (1b)$$

$$\nu = \tfrac{1}{2} - G/(2K) \quad (1c)$$

It is not advisable to use eqn. (1a) to determine ν, because E and G are of the same order of magnitude, and small errors in measurements can therefore result in large errors in Poisson's ratio. The bulk modulus, K, on the other hand, is much larger than the modulus in elasticity or shear, which makes eqns. (1b) and (1c) more suitable for the calculation of ν (the error analysis of eqns. (1a) to (1c) is briefly discussed in the Appendix).

Early measurements of Poisson's ratio in filled elastomers were conducted by monitoring volume changes using gas or hydrostatic dilatometers [*7-9*]. Kruse [*8*] successfully used the Williams, Landel and Ferry (WLF) equation to describe time and temperature effects. Smith and Farris focused on non-linear responses, discussed in terms of binder/filler debonding. Later researchers [*10,11*] measured the bulk modulus in relation to either shear or tensile moduli in order to minimize the effect of experimental errors. Already during the 1960's, Waterman pioneered the use of ultrasonic pulse method in the determination of complex moduli and Poisson's ratio of viscoelastic materials [*12-14,4*]; however, this elegant technique has not seen the widespread use by other researchers that it deserved. More recently, photoelastic techniques [*15*], contact-strain gages [*16*], small-angle x-ray scattering [*17*], and optoelectronic systems [*18*] have been used to measure Poisson's ratio in composites, plastics and elastomers.

Poisson's Ratio for Silicone Sealants

Silicone sealants are an important class of room-temperature-vulcanized (RTV) elastomers. Determining the Poisson's ratio of silicone sealants is an essential task, since these materials are increasingly used as structural adhesives in construction applications, such as structural sealant glazing (SSG) or insulating glass (IG), which frequently require finite-element design calculations. However, published studies of the complex moduli and Poisson's ratio of these sealants are rather scarce.

O'Hara, using the hydrostatic method to determine Young's and bulk moduli of four RTV silicone sealants, found Poisson's ratio in the range of 0.48 to 0.49 for small strains [*19*]. Migwi et al. determined Poisson's ratio of a silicone sealant as a function of temperature, using an experimental technique based on the apparent thermal expansion of constrained specimens. These authors report Poisson's ratio to vary from 0.1 at 50°C to 0.35 at 175°C [*20*]. Ishizaki, Kadono and Miyahara determined the Poisson's ratio of six silicone sealants as a function of strain by analyzing photographs of stressed, flat tensile-test specimens, on which mesh lines had been drawn [*21*]. Figure 2 shows the experimental and calculated Poisson's ratios for the six silicone sealants as a function of tensile strain as reported in their paper. As can be seen, the experimentally determined Poisson's ratios are widely scattered for small strains; however, their average value is close to 0.5. For larger strains (>20%), the Poisson's ratios of the six sealants are very similar, and at large strains (~50%) appear to approach a value around 0.3.

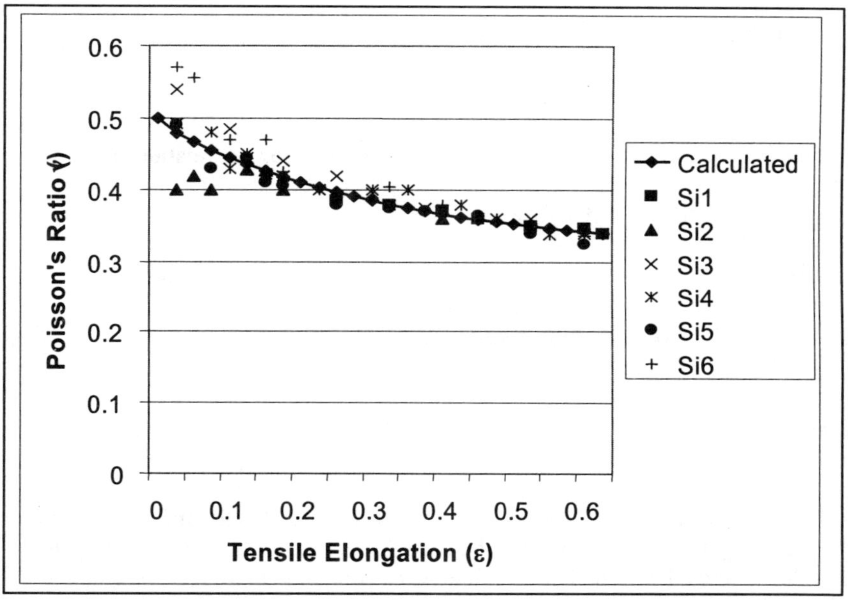

Figure 2 – *Dependency of Poisson's Ratio on Tensile Elongation [21]*

The purpose of this paper is to report on the results of a feasibility study aimed at determining Poisson's ratio of a filled RTV silicone sealant based on simple ultrasonic measurements. It is hoped that this paper sparks interest in this measurement technique and initiates further research into the strain and temperature dependency of Poisson's ratio of silicone sealants.

Experimental

Determination of Poisson's Ratio from Ultrasonic Sound Velocities

For high frequencies (>20,000 Hz), the effective modulus for longitudinal waves, L, and the shear modulus, G, are related to the density of the material, ρ_s, and the squares of the corresponding wave velocities, v_L and v_T [*22*]:

$$L = \rho_s \, v_L^2 \qquad \text{(2a)}$$

$$G = \rho_s v_T^2 \tag{2b}$$

For a homogeneous and isotropic material, the following relationship holds:

$$L = K + 4/3\ G \tag{3}$$

Knowing both L and G, the Poisson's ratio, ν, can then be determined as:

$$\nu = \tfrac{1}{2}\,(L\text{-}2G)/(L\text{-}G) \tag{4}$$

Substituting eqns. (2a) and (2b) into eqn. (4), yields the following relationship:

$$\nu = \tfrac{1}{2}\,(v_L^2 - 2\,v_T^2)/(v_L^2 - v_T^2) \tag{5}$$

By combining eqns. (5) and (1a), Young's modulus can be expressed as:

$$E = \rho_s v_T^2\,(3v_L^2 - 4v_T^2)/(v_L^2 - v_T^2) \tag{6}$$

Equipment Calibration and Experimental Procedure

The basic concepts of back-wall echo sequences, sound path time and attenuation are shown in Figure 3. In the example, the echo display for Material I indicates a high sound velocity and a low sound attenuation, when compared to Material II with low sound velocity and high attenuation.

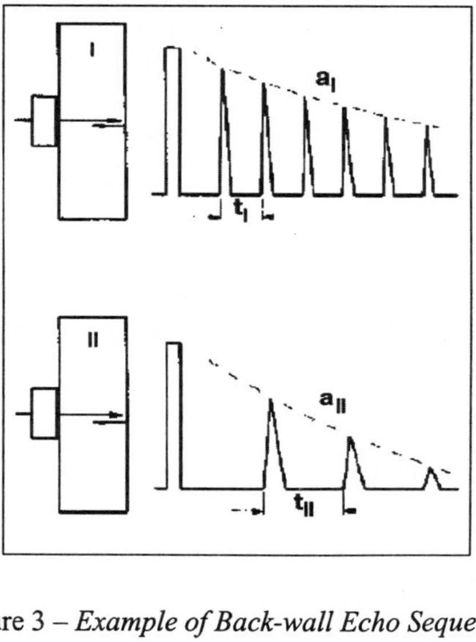

Figure 3 – *Example of Back-wall Echo Sequences*

The apparatus (Krautkramer USD15X and USM22B[2]) consisted of an oscilloscope and an ultrasonic emitter/receiver probe (Krautkramer MK2S[2]). Measurements were performed at frequencies of 0.5, 0.8 and 5 MHz. The equipment was first calibrated by coupling the ultrasonic probe to a block of steel of known thickness and of known ultrasonic propagation speeds. To ensure good contact between probe and sample a coupling fluid (DOW CORNING® 200 Fluid[3], 100 cPs) was used. The resulting back-wall echo sequence on the oscilloscope corresponds to multiples of the path traveled through the object. By adjusting the time base, the distance between two of the echoes was altered to correspond to multiples of the scale graduation on the oscilloscope. This number was then noted. Since the length of the sound path (twice the thickness of the sample) and the speed of propagation were known, the time elapsed between any two of the echo peaks could be calculated and the time corresponding to the graduation on the oscilloscope could be deduced.

Three silicone sealants were tested: a one-part, alkoxy-cure sealant (Si-1) and two two-part, alkoxy-cure sealants (samples Si-2 and Si-3, respectively). All silicone sealants were filled with treated calcium carbonate fillers; the filler content of the three sealant formulations was in the range 40-50% (in weight percent). The cured silicone samples was prepared by drawing out a slab with an area of 65 x 65 mm2 and allowing it to cure for 84 days at standard laboratory conditions (23°C, 55% relative humidity). After completion of the cure period, the thickness of the slabs was about 12 mm. Ultrasound measurements were made to identify void-free sections within the slabs. Within these sections, slab thicknesses for specific ultrasonic measurement points (five locations on the slab) were recorded with an accuracy of ±0.01 mm. Due to the much stronger sound attenuation in silicone versus steel, coupling of the ultrasonic probe to the silicone slab produced a sequence of less than three back-wall echoes. This still allowed calculation of the longitudinal sound velocity, using the time interval represented by each graduation on the oscilloscope and the thickness of the sample at the measurement point as inputs. The velocity of transversal waves could not be measured, even at an ultrasound frequency of 0.5 MHz, due to the high attenuation of these waves. Waterman [*4*] had observed similar high attenuation of transversal ultrasound waves in polyurethane elastomers for temperatures above -20°C.

However, if, in addition to the longitudinal sound velocity v_L, the density, ρ_s, and Young's modulus, E, of a material are also known, the transversal sound velocity can be calculated from eqn. (6). The specific density of the sealants was determined using Archimedes' principle (analogue to the method described in ISO Test Method Determination in Change of Mass and Volume (ISO 10563-1991), but measuring the density directly on the 12 mm thick sealant slabs). Young's modulus of the sealants was determined using a method described earlier [*23*] based on ASTM D412 Test Methods for Vulcanized Rubber and Thermoplastic Rubbers and Thermoplastic Elastomers-Tension (ASTM D 412-98a) with 'dog-bone' type specimens to minimize edge-effects.

---

[2] Krautkramer GmbH & Co. oHG, Robert Bosch Strasse 3, 50354 Huerth (Efferen), Germany.

[3] Dow Corning S.A., Parc Industriel, 7180 Seneffe, Belgium.

Results and Discussion

Using the above method and averaging over the five measurement points, longitudinal ultrasound velocities, v_L, of 1003, 995 and 984 m/s, respectively, were found for the three silicone sealants. From Young's moduli and specific densities, transversal ultrasound velocities, v_T, of 64.4, 88.5 and 90.2 m/s and Poisson's ratios of 0.4979, 0.4960, 0.4957, respectively, were calculated using the "goal seeking" capability of MICROSOFT® Excel[4] computer spreadsheet. Table 1 summarizes the experimental findings, while Table 2 provides on overview of the errors in Poisson's resulting from experimental errors in the measurements of Young's modulus, density and longitudinal velocity (assumed to be ±10%, ±1% and ±1%, respectively). As can be seen, the resulting errors in Poisson's ratio remain rather small (< 0.1%).

Table 1 – *Experimental Findings*

Sealant Type	Young's Modulus (MPa)	Specific Density (kg/m^3)	Longitudinal Velocity (m/sec)	Transversal Velocity (m/sec)	Poisson's Ratio
Si-1 (1 part)	1.8	1420	1003	64.4	0.4979
Si-2 (2 part)	3.3	1380	995	88.53	0.4960
Si-3 (2-part)	3.5	1410	984	90.21	0.4957

Table 2 – *Effect of Experimental Errors on Resulting Error in Poisson's Ratio*

Young's Modulus	Specific Density	Longitudinal Velocity	Poisson's Ratio
±10%			±0.09%
	±1%		±0.009%
		±1%	±0.017%

Summary and Conclusions

In the past, precise Poisson's ratios of elastomers were difficult to obtain, since classical experimental techniques were laborious and fraught with statistical uncertainties. This paper showed that the Poisson's ratios of silicone sealants could be determined with reasonable accuracy (estimated error: < 0.1%) using longitudinal ultrasound velocity and Young's modulus, two parameters that can be measured easily and rapidly. Obtaining accurate values of Poisson's ratio is especially important for nearly incompressible materials, because equations giving the stresses in a body frequently include the term $\nu/(1-2\nu)$. Since values of ν for elastomers are close to 0.5,

[4] Microsoft Corporation, One Microsoft Way, Redmond, WA 98052-6399, USA.

small errors in ν can induce large errors in this term and, therefore, in the predicted stress values.

The paper further showed that the Poisson's ratios of the silicone sealants studied were in the range 0.496 to 0.498, close to the limiting value for incompressible materials. The silicone sealants selected for this study were highly elastic materials with high indentation hardness (40-50° Shore A). Since there is some correlation between bulk modulus and indentation hardness [*19*], this finding does not come as a surprise. However, since silicone sealants can be formulated over a wide range of physical properties, it should not be assumed that silicone sealants in general have Poisson's ratios close to 0.5.

Appendix: Error Analysis for Equations (1a) to (1c)

The errors in estimating the Poisson's ratio, ν, from measurements of the elastic modulus, E, shear modulus, G, and bulk modulus, K, can be derived as follows:

The total differentials of ν can be written as follows:

$$\Delta\nu = \partial\nu/\partial E\ \Delta E + \partial\nu/\partial G\ \Delta G \qquad \text{(a1)}$$

$$\Delta\nu = \partial\nu/\partial E\ \Delta E + \partial\nu/\partial K\ \Delta K \qquad \text{(a2)}$$

$$\Delta\nu = \partial\nu/\partial G\ \Delta G + \partial\nu/\partial K\ \Delta K \qquad \text{(a3)}$$

Applying eqn. (a1) to eqn. (1a) yields:

$$\Delta\nu = 1/(2G)\ \Delta E - E/(2G^2)\ \Delta G \qquad \text{(a4)}$$

which can be written as:

$$\Delta\nu/\nu = -\ (1/\nu + 1)\ \Delta E/E\ +\ (1/\nu + 1)\ \Delta G/G \qquad \text{(a5)}$$

Applying eqn. (a2) to eqn. (1b) yields:

$$\Delta\nu = -\ 1/(6K)\ \Delta E\ +\ E/(6K^2)\ \Delta K \qquad \text{(a6)}$$

which can be written as:

$$\Delta\nu/\nu = -\ (1/(2\nu) - 1)\ \Delta E/E\ +\ (1/(2\nu) - 1)\ \Delta K/K \qquad \text{(a7)}$$

Applying eqn. (a3) to eqn. (1c) yields:

$$\Delta\nu = -\ 1/(2K)\ \Delta G\ +\ G/(2K^2)\ \Delta K \qquad \text{(a8)}$$

which again can be written as:

$$\Delta\nu/\nu = -(1/(2\nu)-1)\,\Delta G/G + (1/(2\nu)-1)\,\Delta K/K \qquad \text{(a9)}$$

In eqns. (a5), (a7) and (a9), the terms $\Delta\nu/\nu$, $\Delta E/E$, $\Delta G/G$, and $\Delta K/K$ represent the errors in ν, E, G and K, respectively.

For example, for a range of Poisson's ratios for elastomers:

$$0.495 < \nu < 0.500 \qquad \text{(a10)}$$

following conclusions can be drawn:

(i) For $\nu = 0.5$, $\Delta\nu/\nu = 0$ in eqns. (a7) and (a9), i.e. there is no error for all possible errors in measuring elastic modulus, E, shear modulus, G, or bulk modulus, K.

(ii) If the true value of ν is 0.495, for example, and supposing that the errors in determining E and K are ±10% and ±20%, respectively, then the error in Poisson's ratio calculated from eqn. (a7) is ±0.3%.

(iii) The same holds true for errors in shear modulus, G, and bulk modulus, K, based on eqn. (a9).

(iv) However, if a similar analysis is performed for eqn. (a5), the error in Poisson's ratio is as high as 90%.

References

[1] Smith, J. C., "The Elastic Constants of a Particulate-Filled Glassy Polymer: Comparison of Experimental Values with Theoretical Predictions," *Polymer Engineering and Science*, Vol. 16, No. 6, 1976, pp. 394–399.

[2] Farber, J. N. and Farris, R. J., "Model for Prediction of the Elastic Response of Reinforced Materials over Wide Ranges of Concentration," *Journal of Applied Polymer Science*, Vol. 34, 1987, pp. 2093–2104.

[3] Holownia, B. P., "Effect of Carbon Black on Poisson's Ratio of Elastomers," *Rubber Chemistry and Technology*, Vol. 48, 1975, pp. 246–253.

[4] Waterman, H. A., "On the Propagation of Elastic Waves through Composite Media (Part II)," *Rheologica Acta*, Vol. 8, No. 1, 1969, pp. 22–38.

[5] Fung, Y. C., *A First Course in Continuum Mechanics*, Prentice-Hall, Inc., New York, 1969.

[6] Gent, A. N. and Hwang, Y.-C., "Elastic Behavior of a Rubber Layer Bonded Between Two Rigid Spheres," *Rubber Chemistry and Technology*, Vol. 61, 1988, pp. 630–638.

[7] Smith, T.L ., "Volume Changes and Dewetting in Glass Bead – Polyvinyl Chloride Elastomeric Composites under Large Deformations," *Transactions of the Society of Rheology*, Vol. 3, 1959, pp. 113–136.

[8] Kruse, R. B., "Time and Temperature Effects," *CPIA Publication*, Vol. 2, 1962, pp. 337–348.

[9] Farris, R. J., "Dilatation of Granular Filled Elastomers Under High Rates of Strain," *Journal of Applied Polymer Science*, Vol. 8, 1964, pp. 25–35.

[10] Rightmire, G. K., "An Experimental Method for Determining Poisson's Ratio of Elastomers," *Transactions of ASME, Journal of Lubrication Technology*, Series F, Vol. 92, 1970, pp. 381–388.

[11] Stanojevic, M. and Lewis, G. K., "A Comparison of Two Test Methods for Determining Elastomer Physical Properties," *Polymer Testing*, Vol. 3, 1983, pp. 193–195.

[12] Waterman, H. A., "Determination of the Complex Moduli of Viscoelastic Materials with the Ultrasonic Pulse Method (Part I)," *Kolloid-Zeitschrift & Zeitschrift für Polymere*, Vol. 192, No. 1-2, 1963, pp. 1–8.

[13] Waterman, H. A., "Determination of the Complex Moduli of Viscoelastic Materials with the Ultrasonic Pulse Method (Part II)," *Kolloid-Zeitschrift & Zeitschrift für Polymere*, Vol. 192, No. 1-2, 1963, pp. 9–16.

[14] Waterman, H. A., "On the Propagation of Elastic Waves through Composite Media (Part I)," *Rheologica Acta*, Vol. 5, No. 2, 1966, pp. 140–148.

[15] Richard, T. G., "The Mechanical Behavior of a Solid Microsphere Filled Composite," *Journal of Composite Materials*, Vol. 9, 1975, pp. 108–113.

[16] Fedors, R. F. and Hong, S. D., "A New Technique for Measuring Poisson's Ratio," *Journal of Polymer Science, Polymer Physics Edition*, Vol. 20, 1982, pp. 777–781.

[17] Dettenmaier, M., "Modern Experimental Methods to Study Mechanical Properties of Polymers," Proceedings of the Fourth Lausanne Polymer Meeting, Ecole Polytechnique Federale de Lausanne, Lausanne, Switzerland, August 28–29, 1989.

[18] Kugler, H. P., Stacer, R. G., and Steimle, C., "Direct Measurement of Poisson's Ratio in Elastomers," *Rubber Chemistry and Technology*, Vol. 63, 1990, pp. 473-487.

[19] O'Hara, G. P., "Mechanical Properties of Silicone Rubber in a Closed Volume," Report ARLCB-TR-83045, SBI-AD-E440224, Large Caliber Weapon Systems Laboratory, Army Armament Research and Development Center, 1983, available from NTIS, Order No. AD-A138129.

[20] Migwi, C. M., Darby, M. I., Wostenholm, G. H., Yates, B., Duffy, R., and Moss, M., "A Method of Determining the Shear Modulus and Poisson's Ratio of Polymer Materials," *Journal of Materials Science*, Vol. 29, No. 13, 1994, pp. 3430-3432.

[21] Ishizakim H., Kadono, M., and Miyahara, T., "Studies on Silicone Sealants for Structural Glazing System, No. 6, Experiments on Poisson's Ratio," (in Japanese), *Summaries of Technical Papers of Annual Meeting*, Architectural Institute of Japan, October 1990.

[22] Lemaitre, J. and Chaboche, J.-L., *Mechanics of Solid Materials*, Cambridge University Press, Cambridge, 1990.

[23] Iker, J. and Wolf, A.T., "Secondary Stresses Induced by Shear Movement in Structural Glazing Sealants," *Materials and Structures*, Vol. 25, 1992, pp. 137-144.

SECTION III

Thomas M. Keranen[1]

When Does it Become a Leak? A Case Study

Reference: Keranen, T.M., **"When Does it Become a Leak? A Case Study,"** *Performance of Exterior Building Walls, ASTM STP 1422*, P.G. Johnson, Ed., ASTM International, West Conshohocken, PA, 2003.

Abstract: This paper explores the basic legal theories under which an owner could seek relief for a leak condition in a legal proceeding; the various standards that could be employed to support the forensic evaluation of the claim, and to "prove" the condition. A case study is used to illustrate the application of these concepts. The paper concludes by making observations and recommendations concerning the process of evaluating and proving a leak condition in a legal proceeding.

---

[1]Principal Attorney and Shareholder; Professional Engineer; Federlein & Keranen, P.C., 6895 Telegraph Road, Bloomfield Hills, MI 48301.

Introduction

Webster's Dictionary defines a leak to be:

> "a crack or opening that permits something to escape from or enter a container or conduit; to permit the escape or passage of something through a breach or flaw."

It may sound a little strange to the reader, especially if the reader is inexperienced with the American legal system, but it is imminently arguable that "improper" moisture infiltration on a structure is not necessarily a legally actionable "leak."

When such a condition exists, the question for the lawyers becomes: When does a "leak" become a "leak" thereby obligating the party responsible for the leak to fix it?

To the non-lawyer, the answer to this question seems to be not only a simple one, but also very obvious. "It rained last night; today there's water standing on the floor; it shouldn't be there; I want it fixed; and, I certainly shouldn't have to pay to fix it."

Obviously, having paid large sums for the design and construction of the building, the owner expects the structure to be serviceable and protected from the elements. At a minimum, the structure should not leak.

But, as the great philosopher, Joe the Bartender might say, "not so fast McGinty, the lawyers ain't done with you yet."

So just what is it that Joe is driving at by his comment? Isn't a leak a leak? Isn't the architectural enclosure supposed to keep the elements out of the building? Didn't the owner buy a waterproof building by hiring the architect to design it and the contractor to build it?

As we have all seen from frequent reports in the various trade journals, "building failures" seem to be a recurring news item. On an industry-wide basis, building failures have grown to be frequent, significant, and expensive legal issues. It is not uncommon to see news reports about structural failures, roof leaks, curtain wall failures, EFIS failures, sick buildings, and the like.

As can also be observed from the news reports, owners having significant financial investment in their facilities are far more aggressive about seeking recovery for their losses when a failure – or perceived failure – occurs. The owner's position is certainly understandable. Typically, the owner did not design or construct the building, but has paid significant sums for those services. As an initial proposition, the owner has the reasonable expectancy that the building will be habitable; that it will be suitable for its intended purpose; and, that it won't leak. However, the owner is exposed to the harm caused by a leak. At a minimum the owner must pay the cost to correct the condition, and suffer any costs of accelerated deterioration of the building resulting from the leak. Further, the owner is also exposed to fixing and/or paying for the damages resulting from the leak, including damage to the building occupants, potentially including property damage, interruption to business and other consequential damages. Depending on the nature of the leak and the use and occupancy of the building, the owner's exposure can often be very substantial.

When a failure event occurs, the owner's typical reaction is to demand an immediate fix from the designer and/or contractor, the parties the owner normally perceives as being responsible. Often the response to the owner's demand is unsatisfactory, usually because of the economic consequences of stepping forward and accepting liability. The owner becomes upset at the apparent evasiveness and seeks legal counsel. Legal counsel seeks advice from the forensic scientist to evaluate the condition; to ascertain the cause of the failure; to ascertain the culpability of the designer and/or contractor; and, to suggest a method for remediating the condition.

Armed with the forensic report, legal counsel pursues a recovery of the owner's losses, taking the matter to the public courts for redress, or to a private alternative dispute resolution mechanism, such as arbitration.

As with any other civil controversy under the American legal system, the basic issues are: Is the defendant (the accused party) liable to the plaintiff (the complaining party) for the failure, and, if so, what are the owner's damages resulting from the failure?

This paper will explore the basic legal theories under which an owner/plaintiff must proceed to present a leak claim in court; the various standards that might govern the resolution of the claim; and, to illustrate the application of these concepts, a case study will be reviewed. This paper will conclude by making observations and recommendations concerning the process for evaluating and proving a leak condition in court.

Elements of a Legally Actionable Claim

In a "civil" action (as opposed to a criminal action where the standards are different), the plaintiff has the burden of proving its case. Burden of proof has been defined as the "necessity or duty of affirmatively proving a fact or facts in dispute on an issue raised between the parties in a cause." Willett v. Rich, 142 Mass. 356, 7 N.E. 776, 56 Am. Rep. 684 (1886). To meet the burden, the proof must be by a preponderance of the evidence, meaning that the "greater weight of the evidence" supports the proposition. In other words, the evidence must be credible and convincing to the mind. This is an important concept, because preponderance does not simply mean there's more evidence in favor of the argument than against it. On the contrary, preponderance embraces the notion that the evidence offered is "superior" and "overbears" the weight of the contrary evidence. Mathes v. Aggler & Musser Seed Co. 178 P. 713; 179 Cal 697 (1919); Barnes v. Phillips, 184 Ind. 415, 111 N.E. 419 (1916).

Whether or not the plaintiff has met its burden of proof is a question for the "trier of fact," which is the jury in a jury trial; the judge in a judge tried case; or, the arbitrator in an arbitration proceeding.

In general, there are two basic theories under which the plaintiff may proceed against the defendant for damages resulting from building leaks:

(1) Breach of contract (which would include contract related theories including express and implied warranties); and,

(2) Negligence (which would include professional negligence (malpractice) in the case of a design professional).

Of course, in order to bring a claim for breach of contract, there must be a contractual relationship between the parties to the dispute. A breach of contract has been defined by our courts to be a "failure, without legal excuse, to perform any promise which forms the whole or part of a contract." Friedman v. Katzner, 139 Md. 195; 114 A. 884, 886 (1921). Under a breach of contract theory, the plaintiff must demonstrate by a preponderance of the evidence that the defendant has committed a breach of a material condition of the contract; that the plaintiff has been damaged by the breach; and, that the plaintiff's damages are directly related to, and are the result of, the breach.

In the construction industry, contracts to build a building are normally voluminous in nature, and include the drawings, specifications, general conditions and other elements of the "contract documents." While many of the elements of the typical construction contract are technical in nature (such as the technical specifications), it is important to recognize that they are indeed contract requirements, and they have legal significance. In fact, there is very little contained in the construction contract that does not have legal significance. Typically, the construction contract is drafted by the architect or engineer, using standard industry forms, specifications from previous jobs, and custom-drafted provisions for special elements of the project. The lawyer is seldom involved at this stage.[2] The typical contract, because of the manner in which it is prepared, is rife with conflicting provisions, poorly drafted wording, contradictions, and the like, all of which makes interpretation of its requirements a formidable task.

Taken in the light of the plaintiff's burden of proving its case, the relative complexity of "making a case" becomes readily apparent. Simply stated, it's a difficult thing, even in the simplest of cases.

The task has become even harder because of the advance of technology, as well as the increasingly defensive nature in which construction contracts are being drafted. The industry is seeing more and more exculpatory language included in contract documents intended to shift the risk, usually to the contractor.

There was a time past when construction specifications and drawing "details" were very specific and comprehensive. A look at drawings and specifications for projects from the early 1900s illustrates this point. Building the project was rather "cookbook like", in that all that the contractor had to do was to construct the project in the manner shown and specified by the design professional. Under traditional methods of project

[2] ...indeed, for good reason. As one industry wag remarked, if a lawyer drafted the contract it would be so heavy that you'd need a crane to lift it and so onerous that no one could afford to build it!

delivery the lines of responsibility were likewise clear. It was clear where the architect's duties started and stopped. Similarly, it was clear what the contractor's duties were.

In more recent times, as technology has advanced (and as design professionals became more aware of their exposure to liability for defects in their designs and more aggressive about protecting their exposure with contract language), it is now more commonplace for construction contracts to include specifications that are "design-build" or "performance" orientated. These specifications require the contractor to actually perform a portion of the design for a particular building system in order to meet specified performance criteria.

By reference, performance specifications often adopt established performance criteria, as defined by recognized industry groups such as ASTM, for the purpose of defining the acceptability of the contractor's performance. Examples in construction contracts are commonplace. One such example is as follows (as it pertains to wood windows and doors on a specific project):

Performance Requirements

A. Window units shall meet Grade 60 specifications in accordance with NWWDA 1.S.-2, except where more stringent requirements are specified otherwise.

. . . .

D. Air Infiltration:

1. Window unit air leakage, when tested in accordance with ASTM E283, at 1.57 psf (25 mph), must be 0.03 cfm/ft of crack or less.

Since the contractor, particularly in current times, is usually not an expert in the design or construction of the technical systems specified (nor usually is the design professional for that matter), the actual design or "applications engineering" of the specified system is often performed by the manufacturer or distributor of the system. Often the manufacturer or distributor is a subcontractor or supplier to the contractor, and may well furnish and install the system.

This delegation of design responsibility makes some practical sense, as who would know a system better than the manufacturer who has developed and manufactured it.

Unfortunately for the legal system, the dilution and erosion of traditional lines of liability has exponentially complicated the proofs of a claim. Where does design now stop and construction start? Does the contractor have inherent design responsibility for design-build and performance specifications? What about the Designer of Record for the whole project? What liability does the architect have if the architect approves shop drawings and submittals for a system the architect does not have expertise in? The questions are endless.

In the end, if there is a failure in the specified system, the only questions that matters are:

(1) Who is responsible for the failure, the architect for designing and/or specifying it, or the contractor for designing and/or building it?

(2) Has there been a breach of the contract (either the architect's; the contractor's; or both?)

The situation becomes even more difficult when notions of professional negligence (or malpractice) are considered.

In concept, negligence is a "tort" which is a separate cause of action independent of the breach of contract theory. Courts have defined a tort to be a private or civil wrong or injury, independent of contract. It is a violation of duty imposed by civil law upon all persons occupying the relation to each other which is involved in a given transaction. Coleman v. California Yearly Meeting of Friends of Church, 27 Cal.App.2d 579, 81 P.2d 469 (1938).

Professional negligence is very similar in concept to ordinary negligence. Professional negligence, commonly called malpractice, is negligence by a professional in the performance of professional duties. Who is a Professional is normally defined under state licensing laws and typically includes doctors, lawyers, architects, engineers, and other professionals.

In the case of design professionals (which applies equally to both architects and engineers), under the theory of malpractice, the plaintiff must prove – by the preponderance of the evidence – that: 1) the design professional owed a professional duty to the plaintiff (the plaintiff does not necessarily have to be the client); 2) the design professional breached that duty by violating the professional "standard of care" applicable to the profession; 3) the plaintiff has been damaged by the design professional's breach of duty; and, 4) that plaintiff's damages are the foreseeable and consequential result of the breach of duty (i.e., that the design professional's breach was the proximate – or legal –cause of plaintiff's damages).

In recognition that design is a combination of science and art and is not perfect or exact, the courts have consistently defined the professional standard of care as being that level of care exercised by a similarly situated design professional using ordinary care. While the "bar" may be raised because of the particular level of expertise held by a specific design professional (for example, an expert in a particular kind of design) the significance of this legal definition is that a design error or mistake is not automatically legally actionable. As long as the mistake was not outside the prevailing standard of care, the designer will enjoy freedom from liability for the mistake. The design error only becomes actionable when the mistake violates the prevailing standard of care – specifically, when a similarly situated designer using ordinary care would not have made the mistake. Tiffany v. Christman Company, 287 N.W.2d 199; 93 Mich.App. 267 (1979).

The concept of malpractice is frequently misunderstood by the public, which usually expects a higher level of accountability from the design community. The concept lacks objective definition, and it is difficult to understand why the designer might be excuse from responsibility for defects in the project. The concept is probably understood only slightly more by lawyers and designers in the business, and it provides little guidance to the industry for determining when malpractice has actually been committed.

When trying to prove malpractice in a litigation setting, the definition leads directly to the "battle of the experts" in the courtroom. To prove malpractice, plaintiff must present "expert" testimony by a design expert that the conduct complained of failed to meet the professional standard of care. In defense, the defendant presents rebuttal expert testimony that the conduct met the standard of care.

For judges and juries this presents difficulty because, typically, neither have the technical expertise to properly evaluate the competing expert opinions. Often the result is a poorly reasoned decision that defies technical logic – the so-called "what were they thinking" decision.

The explanation for questionable decisions is often that the judge or jury could not understand the technical arguments, and instead relied upon the testimony from the expert they thought was best. In the end, the decision was not necessarily based upon the expert's technical competence or analysis, but rather was the result of the expert's presentation and communication skills. The jury simply liked one expert better than the other and adopted the opinion of the expert they liked best.

For the plaintiff, the situation can mean disaster. Not only is the plaintiff left with the mess caused by the leak in the first place, the plaintiff has also lost the lawsuit and is now left without recourse.

Arbitrations can sometimes lead to better results. Theoretically, at least, arbitrators are selected by the parties because they hold expertise in the subject matter of the dispute. Thus, presumptively, the arbitrator is better suited than a judge or jury to hear the technical evidence and properly evaluate it without being overly influenced by a particular expert's presentation skills.

What this dialog is meant to illustrate is that, in court, there are no sure things even when the answer appears obvious. One observer has opined that when it comes to construction disputes, our system of jurisprudence is ineffective because it takes too long, costs too much, and produces results that are far too unpredictable. We might well find after the conclusion of the trial that a leak may not be a leak within the meaning of legal definitions and the burdens of proof, even as water is pouring in. Unfortunately, that decision becomes the law of the case and is binding on the parties.

This dialog is further meant to illustrate the need within the construction industry for clear and objective forensic standards for the purpose of evaluating leak issues, and which are capable of better understanding by judges, juries and arbitrators.

Testing Procedures as Contract Performance Standards

As we have seen above, many construction contracts adopt performance standards established by various industry groups and associations, such as ASTM, by reference.

One of the primary problems with adopting standards of this kind to measure contractor performance is that the standards were not necessarily developed for that purpose. In most cases, the standards were not meant to test the contractor's compliance with contract specifications, but rather to test products during their manufacture. Many times, the standard referenced by the contract is not applicable to the system being constructed in the field. Thus, in the context of proving liability, we see a misapplication by the design professional of industry standards to the specified work. Where the specified standard might be relevant to the work, it is often difficult to perform the specified tests on the assembly in the field, which usually requires the employment of a testing laboratory or specialized apparatus. This can be expensive and time consuming, and does not always provide a reliable standard by which to measure performance.

As an example, even ASTM standards have drawbacks when used for forensic purposes. The ASTM standards tend to address a specific item or element of an assembly, such as shingles or insulation, and they do not usually address the assembly itself, such as the roof or curtain wall.

An illustrating example would be a curtain wall. A curtain wall is a non-load bearing exterior wall that is "hung" on the building's frame. The curtain wall may contain steel studs, sheathing, insulation, a vapor barrier, windows and siding. ASTM has standards to test each element comprising the curtain wall, but not necessarily the curtain wall itself. Thus, we are left with standards that address the specific elements of the curtain wall, but not the overall assembly. While this may not pose a problem for the manufacturing industry, it certainly poses a problem for an architect trying to write specifications; a contractor trying to construct the project; and, a plaintiff seeking to prove, by the preponderance of the evidence, that an architect has committed malpractice or a contractor has breached a contract specification resulting in a leak.

This limitation frequently forces the forensic evaluator to exercise some independent judgment relative to the application of the standard or the methodology by which the standard is applied. This, of course, exposes the evaluator to a claim of being subjective by the party who doesn't like the results of the test. The argument can be made that: 1) the recognized test procedure was not followed, thus, the results of the test should not be considered; or, 2) that the wrong test procedure was applied, thus, the results should not be considered.

Another significant problem when employing standards is that any standard has some tolerance or latitude for performance built into it. Usually, the standard provides some range of acceptability before the performance is considered to be a failure. The standard necessarily has to be this way, as no performance is "perfect" every time.

The application of a tolerance is legally significant as well. The law typically obligates the contractor to substantially comply with the contract requirements. Substantial compliance contemplates a performance that is less than strict compliance, with some deviation permitted from every last requirement of the specifications, so long as the deviation is not material. One court has defined substantial performance as being that "...level of completion where construction has progressed to the point that the building can be put to the use for which it was intended, even though comparatively minor items remain to be furnished or performed in order to conform to the plans and specifications of the completed building." Southwest Engineering Co. v. Reorganized School District R-9, 434 S.W. 2d 743, 751 (Mo. Ct. App. 1968). Thus, not only do the standards allow some tolerance or variation in an acceptable performance, but the legal concept of substantial performance does as well.

It should not be surprising that it is difficult for the plaintiff to make its case where objective standards are lacking or compromised to prove a failure to comply with specifications (a breach of contract), or that the design was defective (malpractice).

An Illustrative Case

A case illustrating these issues involves a multi-story, upper quality hotel containing approximately 500 guestrooms. The hotel was constructed in the mid 1990s in the Midwest.

The exterior walls of the hotel were designed as a curtain wall in a typical manner. In general, the exterior walls consist of: steel studs; insulation; sheathing; weather resistive barrier; lap and flat panel siding; operable windows and, in certain locations on lower elevations, manufactured stone veneer.

The exterior cladding was designed to function as a water-shedding element with a weather resistive barrier system beneath it. The weather resistive barrier was to be constructed of asphalt saturated felt, supplemented with self-adhering membranes and flashing. The weather resistive barrier occurs beneath the exposed exterior water shedding cladding, irrespective of whether it is lap or flat panel siding or manufactured stone. The weather resistive barrier, to adequately maintain its integrity, has to be integrated with other elements of the structure passing through it such as the building's windows, doors, exhaust fans, and structural elements.

It is readily apparent that for such an assembly to work properly to shed water, it must be carefully installed to achieve its planned function. There can be no breaks or openings in the barrier, or moisture could penetrate at those locations.

In order to expedite construction of the hotel, the construction of the exterior walls was "panelized." The architect prepared the drawings and specifications for the panalization of the curtain wall. The contractor was obligated to follow the drawings and specifications when building the walls. The concept of panelizing the curtain wall was adopted to standardize the wall sections where possible, and to allow the panels to be assembled under shop (or controlled) conditions. Once assembled, the panels were to be hoisted into location and attached to the structural framing. While the use of panels increased the production on the project (because the panels could be constructed during inclement weather), their use did result in a long vertical joint at either edge of the panels as they were stacked one on top of the other over the multiple floors. It was the design intent that the vertical joint be caulked in order to make it weather tight.

Because of design changes while the panels were in fabrication, altering many of the dimensions, window locations, "bump outs" and other features, the panels could not be standardized to the extent originally anticipated. Furthermore, the panels could not be built to a completed stage (with siding attached) in the fabrication shop as planned.

Instead, the panels were fabricated without all of the sheathing or siding in place, and were erected partially finished. The balance of the sheathing and most of the siding was installed after the panels had been erected.

After construction of the exterior walls was finished, and while the interior work was proceeding, construction crews noticed water infiltration in unexpected locations throughout the building. Water leaks are not unusual during construction and, initially, there was no particular alarm taken as a result of the leaks. Crews investigated the leaks and repaired obvious sources of leaking, such as missing flashing, siding or caulking.

The leaking continued, however, with leaks reported in random areas of the building following moderate and severe weather events. The most severe ingress was reported on the first and second floors on the north and west elevations. Ingress was severe enough to necessitate replacement of significant sections of drywall, wall coverings, and trimwork.

Often, however, the manifestation of a leak was not "big," meaning that a large volume of water was not necessarily observed. Many times, the leak was evidenced by a relatively small stain on the wall, or a small quantity of standing water on the windowsill or floor.

A further difficulty manifested by the leaking was that it was intermittent and inconsistent in terms of location or quantity. Sometimes, leaks would appear in the same locations, sometimes not. Sometimes the leak was large (a big puddle) and sometimes it was small. Sometimes the leak would appear after a storm, but not reappear after the next storm, only to reappear later after a third or fourth storm.

Many of the leaks manifested themselves around the windows in the building.

In general, the windows were a "window assembly," consisting of a group of individual fixed, awning and casement windows joined together to form the architectural design. The windows were commercial brand wood windows with exterior aluminum cladding.

The leakage presenting in the vicinity of the windows generally manifested itself by staining on the interior wall at the upper corners of the window heads and dampness on the interior frames and sash, with accumulations in the lower tracks and on the sills. Sometimes there was dampness on the floor beneath the windows, as well as at locations along the exterior wall that was not beneath the windows.

The interior walls are drywall, with the final finish being either paint or vinyl wall covering. The final floor finish was carpet.

At one point, after becoming frustrated because the leaks could not be stopped, a portion of the wall was opened to accommodate further investigation. Nothing of significance was revealed.

After the owner occupied the hotel, it became more difficult to observe the leaks because the guests' use of the rooms limited the owner's ability to inspect them and further provided alternative explanations for finding moisture in the rooms.

The only meaningful leak observations were the reports of the hotel service and maintenance staff. Unfortunately, findings were intermittent and not necessarily reliable.

First, the finishes tended to hide the leaks as the moisture was either absorbed by the carpeting, or would run down the vinyl wall covering without leaving a stain. Thus, many of the leaks may well have gone undetected. In addition, there were many possible explanations, beyond a leak, for some water being in the rooms. If the carpet was observed to be wet and the maid happened to observe it, the water may have been the result of a window being left open, something being spilled, or someone walking across the carpet after showering while still wet. If the sill was wet, the window may have been left open during a rain.

Ultimately, the owner became convinced that the building was leaking where it was not supposed to, and that the cause of the leaking was not obvious. The owner placed the architect and contractor on notice of the condition, and sought the advice of a forensic investigator who had significant experience in similar situations.

The investigator reviewed the design documents, construction practices, and construction information pertaining to the design and construction of the curtain wall. Subsequently, field tests were performed at two locations where consistent leaking had been reported and at one location randomly selected. With the building being in use, it was difficult to arrange the logistics for more sampling without creating disruption to the hotel guests.

The field tests consisted of the controlled dismantlement of selected components of the exterior wall system in selected locations. Materials were observed in their as-built condition; materials were sampled, and water tests were performed. Field tests were performed at two locations where leaking had been consistently reported and in one location randomly selected. Some of the materials (the windows, for example) were tested at the laboratory.

The forensic evaluation revealed a number of construction defects resulting in breaches of the barrier system; as well as design defects consisting of questionable and inadequate design strategies to create an appropriate moisture barrier; uncoordinated combinations of wall building products and materials in an inappropriate manner; failure to employ the use of recognized wall "systems"; poor or insufficient detailing, and the like.

The field observations suggested the barrier system had been compromised by design and construction defects, and moisture was migrating into the wall as a result. It was believed that most of the moisture was staying inside the stud spaces and other cavities within the wall, with only a fraction of the moisture making its way to the interior of the building, which required the water to move laterally. Lateral movement could only happen when the vertical path of the moisture was obstructed by a framing member, a penetration, or a window assembly.

There was also evidence of leakage in certain windows within the window assembly, particularly the fixed sash windows. The fixed sash window units had been manufactured as operable sash, but had the operating mechanisms removed to make them fixed. Plastic clips were installed inside the frames, apparently to hold the units square during shipping, with "remove" labels affixed to the clips. However, because the operating mechanisms had been removed, the shipping clips could not be removed in the field without disassembling the window frame. Because there was no practical method by which the clips could be removed, they were left in place during installation. The effect of leaving the clips in place, in combination with the removal of the closing mechanisms prevented the windows from being drawn (latched) tight against the weather seal around the perimeter.

With preliminary findings in hand, the owner placed a demand upon the architect and contractor to remediate the condition. Both refused.

The contractor refused the owner's demand claiming: 1) that the owner had not proven that the building actually leaked; 2) that if the building did leak, the leaks were caused by poor and defective design, not defects in construction; and, 3) that the opinions of the forensic evaluator should not be considered because the methods employed were inconsistent with industry procedures and standards.

The owner, confronted with a leaking building and the suggestion that the leaks were caused by both design and construction defects, sought relief by demanding arbitration against both the architect and contractor.

Under the rules of arbitration, each case had to remain separate and could not be consolidated.

The case against the contractor moved forward, ahead of the case against the architect, and will be examined here as it presents the best illustration of the legal considerations identified above.

One of the contractor's primary defenses was that the owner failed to prove (by legal standards) that the building leaked. Citing the large number of possible explanations for moisture penetration, and the rather poor documentation of actual leak occurrence, the contractor claimed that the evidence offered by the owner failed to demonstrate that the building was actually leaking, let alone that any leaks were related to something done by the contractor. To support its position, the contractor pointed to a location where an obvious construction defect existed (a location where the butt joint in the plank siding exceeded the recommended tolerance for example), but where there was no leak.

The contractor's most remarkable defense in this regard pertained to the window leaks. The forensic evaluator concluded that some of the leaking could be explained by the fixed sash windows not being drawn tight against the weather seal. The theory was that during a wind driven rain, moisture could penetrate between the seal and the frame since the frame could not be latched normally (because of the absence of the latching mechanism and the presence of the clips).

To support the theory, lab testing was performed on a sample window assembly. Individual windows were assembled to model one of the window assemblies installed on the building. The assembly was then tested in the lab under ASTM E 283-91 for the purpose of determining whether or not the assembly met the air infiltration requirements of the contract (see the specification cited on page 4 above). The assembly was also tested under ASTM E 547 to determine its resistance to water penetration. Because of the size of the assembly, it was not possible to install gages to actually measure the volume of air infiltrating the assembly, so the air infiltration testing consisted of a "smoke test" under controlled pressure. The moisture penetration testing was conducted under controlled pressure and water volume on the exterior side of the window, but no gages were available to measure to volume of water penetration. The testing was demonstrated in the presence of the arbitrators during the arbitration proceeding.

The smoke test demonstrated obvious air infiltration under relatively low pressure, but without quantified results. The water penetration test demonstrated obvious water infiltration under low pressures, coincidently at the same locations on the window frames that actual leaks had been observed in the building.

The contractor objected to this testing because the test procedures actually employed did not strictly comply with the ASTM testing standards. The contractor further argued that the ASTM standards used did not apply to testing an assembly of windows, and were intended for testing single windows only. For these reasons, the contractor argued that the test results should not be considered by the arbitrators as probative evidence of any window leaks.

Furthermore, the contractor contended that any observable moisture penetration of the window frames, including any accumulation on the windowsill, was authorized by industry standards and was not a leak.[3] The contractor pointed out that, under the test standards (assuming that the test was properly performed), some moisture penetration is permitted and the window unit does not fail the test, unless the accumulated water becomes uncontrolled: to wit, not until the accumulated water spills over the edge of the sill and runs down the wall. Since the water did not spill over the sill during the test, there was no window failure – despite the presence of water on the sill. Thus, even though water could be observed standing on the sill, the window did not leak within the meaning of the test standard. Hence, the leak was not a leak.

The contractor also criticized the evaluator's field-testing procedure. For example, during the field test, the investigator flooded the wall at one of the test locations (after the interior drywall had been removed) to see if the wall could be made to leak. It was not intended that the flooding be a "test" under some prescribed test method. On the contrary, it was the evaluator's thinking that if the wall or window could be made to leak, the path of the ingress could be observed which would help better understand the nature of the leaking – where the water came from and where it was going. The wall did leak during this exercise, and the investigator obtained valuable information concerning possible sources and paths of the ingress. The contractor objected to this method claiming that it failed to comply with any industry standard or ASTM test method, and that it should not be allowed as probative evidence that the wall or window leaked.

Thus, concluding, the contractor argued that the owner failed to meet its burden of proof in that no evidence was presented, let alone a preponderance of evidence: 1) that the building leaked; 2) that the contractor had breached a material requirement of the contract in a manner to cause any leaking; 3) that the contractor was negligent in any way; 4) that the owner had been damaged by any leak (if there are no leaks, how can

[3]See Paragraph 3.2.3, water penetration, ASTM E 331 and Paragraph 3.2.3, water penetration, ASTM E 547.

there be damage?); or 5) that any leak or any damage claimed was linked to or caused by any action or inaction by the contractor.

Without a detailed explanation as is typical of arbitration awards (thus, it is not possible to ascertain their reasoning), the arbitrators determined that the contractor was not legally responsible to the owner for the leaks in the hotel. The obvious conclusion is that the arbitrators accepted one or more of the contractor's arguments.

Commentary on the Award

To a person who has little experience with the legal system, the results of the owner arbitration with the contractor are unbelievable. How could the arbitrators rule in such a manner? Isn't a leak a leak? The award was certainly unexpected to the owner, who was convinced the building leaked, and that either the contractor or architect (or possibly both) was responsible for it. The owner knew for sure that it did not cause the leaks, having done no design or construction, and merely having paid the bills for those services. Further, the owner did not believe that it had authorized either the contractor or architect to give it a building that leaked. At the risk of a bit of understatement, suffice to say that the owner was bitterly disappointed by the arbitrator's award.

From the legal perspective, the possibility of such an award is more understandable. As indicated earlier, no case is easy, especially construction cases.

The legal reasoning for the award goes directly to the owner's obligation to prove the material elements of its case by a preponderance of the evidence, and the realities of the situation. Obviously, at the time the leaks were discovered, the owner's primary concern was getting the building finished and putting it into service, not documenting a court case. It must be further recognized that the leaks were not thought to be significant initially, and it was believed that explanations for the leaks could be found and fixes made as construction was completed. However, by the time it became clear that the leaking was significant, and without obvious cause, the building was occupied thereby making further exploration and documentation difficult. Destructive testing was limited because the owner was reluctant to take major areas of the building out of service, was very reluctant to disturb guests and, was understandably very reluctant to make a public display of the problem. As described above, information from the hotel's service staff was limited and was certainly not "scientific" in nature. By the time the forensic evaluators arrived, they were confronted with a difficult situation to analyze, and with limited diagnostic resources at their disposal. The intermittent and largely indiscernible nature of the leakage further complicated the evaluation.

The contractor seized upon these shortcomings in the owner's case, and used them in conjunction with its legal arguments. For example, the contractor argued that it had substantially complied with specifications. The contractor claimed that the specifications misapplied certain test standards (for example, that ASTM E 283 does not apply to a window assembly and that the assemblies should not have been tested under that standard); and, that the results of the questionable testing practices employed by evaluators actually yielded a "no defect" conclusion when applying the very standards specified in the contract.

The arbitrators obviously found that the owner failed to prove its case against the contractor by the preponderance of the evidence. In any material regard, the award, as rendered, was the logical legal result.

Conclusions and Recommendations

The obvious conclusion from this result is: winning a recovery for building leaks is not automatic, nor is it an easy thing to do. Showing that water penetrates the building's enclosure is simply not enough. To prevail, the owner must be able to prove every element of its case by a preponderance of the evidence.

Using industry standards, such as those established by ASTM, can be helpful in meeting this burden; however, there are limitations that must be recognized. The standards typically address individual construction components and not the assemblies normally being constructed. Furthermore, the standards allow for various performance tolerances that are often inconsistent with the owner's expectations.

It would be helpful (to the lawyers, at least) if industry groups, such as ASTM, would establish standards that better address forensic needs that could function to establish baseline, quantified, performance criteria for assembles commonly used in construction.

With the establishment of better performance criteria, the risk of "what were they thinking" court decisions could be reduced.

Daniel J. Wise,[1] Bipin V. Shah,[2] Dragan Curcija,[3] and Jeff Baker[4]

Evaluation of the Condensation Index Rating as Determined using the Proposed Testing Method in the NFRC 500 Draft Procedure

Reference: Wise, D. J., Shah, B.V., Curcija, D., and Baker, J., **"Evaluation of the Condensation Index Rating as Determined Using the Proposed Testing Method in the NFRC 500 Draft Procedure,"** ***Performance of Exterior Building Walls, ASTM STP 1422***, P.G. Johnson, Ed., ASTM International, West Conshohocken, PA, 2003.

Abstract: This paper presents a comprehensive and detailed assessment of the proposed NFRC 500 Procedure currently under development in the NFRC Technical Committee, as developed by the NFRC Condensation Subcommittee. The proposed NFRC 500 Procedure contains a new element in the evaluation of condensation, that being the utilization of computer software in analyzing and calculating the condensation index for comparative purposes. For those products that can't be simulated, a test only option is also offered.

The primary focus for this paper is the analysis of the testing portion of the proposed NFRC 500 Procedure, emphasizing the testing calculations as defined in the procedure. Two NFRC annual round robins will be characterized, 1999 and 2000. The 1999 and 2000 NFRC round robin tests employed a nominal 1520 mm x 910 mm (60" x 36") non-thermal aluminum horizontal sliding window with high performance glazing. The simulated results will be used as a benchmark for the analysis of the testing data. Each round robin test specimen was tested at the NFRC-accredited Testing Laboratories. The tested fenestration thermal performance ratings (U-factors) were acquired in accordance with *NFRC 100 (1997): Procedures for Determining Fenestration Product U-factors* and *NFRC Test Procedure for Measuring the Steady-State Thermal Transmittance of Fenestration Systems (April 1997).*

This paper details specific data regarding various measured and calculated performance values as required by the NFRC Procedures and Program Documents.

[1]Program Director, National Fenestration Rating Council, 1300 Spring Street, Suite 500, Silver Spring, MD 20910.

[2]Program Manager, Engineering, National Fenestration Rating Council, 1300 Spring Street, Suite 500, Silver Spring, MD 20910.

[3]Center for Energy Efficiency & Renewable Energy Engineering Lab Building, University of Massachusetts, Amherst, MA.

[4]WestLab, 24-94 Bridgeport Road E., Suite 311, Waterloo, Ontario N2J 2J9.

Keywords:

Condensation Index, Coefficient of Variance, Round Robin, National Fenestration Rating Council (NFRC), Dew Point Temperature, Reproducibility

Overview

The National Fenestration Rating Council (NFRC) draft procedure, *NFRC 500* [1], provides two methods of determining a condensation index rating for fenestration products, including windows, entrance doors, sliding glass doors, skylights, door-lite, and curtain wall. Those methods are simulation and physical testing. Fenestration products include residential, commercial and site-built applications. For the purposes of rating, the product is modeled and rated at a net zero air leakage, meaning that product is sealed and air leakage effects on condensation index rating are not considered.

The total product is evaluated for condensation, and the rating is a total product rating. Manufacturers can now, effectively and efficiently, obtain condensation index ratings for their fenestration products by using software tools developed to model and calculate a Condensation Index (CI) rating in concurrence with U-factor, Solar Heat Gain Coefficient (SHGC) and Visible Transmittance (VT). Software programs are reviewed by NFRC. If the software meets criteria established by NFRC, it is then used to determine thermal performance ratings. In essence, four ratings can be obtained simultaneously by using simulation software.

Definitions

The following definitions are referenced from
Fundamentals of Heat and Mass Transfer; Incropera. F.P. and DeWitt.

Conduction - the method by which heat is transferred due to free valence electrons moving through the metallic crystalline lattice or due to agitation of atoms vibrating about their equilibrium points in the lattice.

Convection - the method by which heat is transferred by the bulk, or macroscopic, motion of the fluid.

Radiation - the heat transferred process as a consequence of energy-carrying electromagnetic waves, emitted by atoms and molecules resulting from changes in their energy content. Simply, energy transferred due to propagation of electromagnetic waves.

Ambient Temperature - Temperature at a given set of environmental conditions. For condensation resistance, the surrounding localized air temperatures would be considered the ambient air temperatures.

Relative Humidity - the ratio of the amount of water vapor in the air compared to the maximum amount of water vapor that the air could hold at that particular temperature. When the air is holding all of the moisture possible at a particular temperature, the air is said to be saturated.

Dew Point Temperature - the temperature to which air would have to be cooled for saturation to occur. On the surface of the window, this is the surface temperature at which condensation would first begin to form when the surface temperature, for given relative humidity conditions, are at or below the dew point temperature. If conditions are such that condensation forms when the surface temperature is above 0C (32F), condensation would be in the form of water droplets; and if the temperature is at or below 0C (32 F), the condensation would be in the form of frost or ice.

Physical Testing for Condensation Index

Physical testing is to be used only in the case where a product cannot be accurately simulated using currently approved NFRC software tools. In this instance, physical testing is the only alternative offered to obtain condensation index ratings. The test can be performed simultaneously with the U-factor testing using the *NFRC Test Procedure for Measuring the Steady-State Thermal Transmittance of Fenestration Systems.* For condensation evaluation, thermocouples are attached to the interior surface of the test specimen in pre-specified locations.

Standardized NFRC conditions are used for the evaluation and rating of the condensation index. This is true for either the simulation method or the test method as specified in *NFRC 500.*

Calculation of Condensation Index (CI)

The determination of a condensation index rating for a particular product is performed using the same calculational methods, whether the simulation or test only method is used. When the computer modeling is performed with NFRC-approved software, the temperature measurement at a locations, measurement of areas at or below the dew-point temperature, and CI calculations are an automatic routine of the software. If the test only method is used, the same CI calculation principles apply. However, actual thermocouple sensors are placed on the interior of the product. Actual temperature readings are then obtained during testing, and these measured temperatures are evaluated and placed in the equations to calculate the condensation index of the frame and the condensation index of the glass.

To determine the condensation indices, record the average interior ambient surface temperature for each individual thermocouple location for testing. The next step is to calculate the wetted area assigned to each individual surface thermocouple sensor as described in NFRC 500, Section 6.2.2. Next, the percentage areas are determined by considering the preassigned area for each individual centerline thermocouple locations. Continuing, identify the thermocouple temperatures that are less than the dew point temperatures and calculate the frame areas and glazing areas that have surface temperatures at or below the three prescribed dew point temperatures at 30%, 50% and 70% relative humidities. Finally, determine the condensation index of the frame, CI_f, and of the glazing, CI_g by using the equations as defined in *NFRC 500*.

NFRC 500 uses three defined relative humidities (30%, 50% and 70%) for evaluation and rating purposes. First, each specified relative humidity level is evaluated separately, and then, by using mathematical formulas to assess the performance at each relative humidity, are combined to provide a total product condensation index rating, which is the lower of CI_f or CI_g.

Ratings and Sizes

NFRC provides a fair, accurate, and credible rating system for thermal performance properties such as U-factor, Solar Heat Gain Coefficient (SHGC), Visible Transmittance (VT) and Condensation Index (CI). Size is an important characteristic of the product when considering comparison of these ratings from product to product. Size of the product is important when these ratings are considered because of the area ratios of the frame, edge-of-glass, and center-of-glass. However, for a condensation index, there are two areas that are considered: the glazing area (comprised of the center-of-glass and the edge-of-glass) and the frame area (frame and sash components). Differing sizes of the same product may have an effect on the ratings.

NFRC has addressed the size issue for each operator type and provides referenced sizes for comparison purposes. The current sizes can be found in *NFRC 100*, Table 1. Proposed revisions to the sizes are currently being considered, which would basically provide standard operator sizes for the following general categories: windows, entrance doors, double doors/glazed wall systems/sloped glazing, and sidelites/transoms.

Analysis of the 1999 and 2000 Test Round Robin

The proposed condensation index equations, as found in the NFRC 500 proposed document, have been used to analyze the non-thermal aluminum horizontal sliding window with high performance glazing. Two identical products were shipped to each NFRC-accredited Testing Laboratory, one for 1999 and one for 2000, as required in the NFRC Laboratory Accreditation Program.

The following tables illustrate the results of the 1999 and 2000 round robins, along with the combined results. See Figures 1, 2 and 3 in the Appendix for graphical representation of the results. Figure 1 is the 2000 test results; Figure 2 the 1999 test results; and Figure 3 is a combined 1999/2000 test results for the Condensation Index rating.

Table 1 - *Condensation Index Values*

Laboratory	1999 CI_f	1999 CI_g	2000 CI_f	2000 CI_g	Averaged Specimen CI_f	Averaged Specimen CI_g
1	20.6	52.2	18.5	54.5	19.6	53.4
2	17.8	53.3	21.6	50.0	19.7	51.7
3	19.7	48.0	18.3	50.4	19.0	49.2
4	19.2	53.6	20.4	52.1	19.8	52.9
5	22.0	57.6	28.1	59.9	25.0	58.8
6	17.4	51.8	15.7	50.9	16.6	51.4
7	21.1	52.7	17.0	54.2	19.0	53.4
8	20.9	53.8	19.4	53.6	20.2	53.7
Average:	19.8	52.9	19.9	53.2	19.9	53.0
High:	22.0	57.6	28.1	59.9	25.0	58.8
Low:	17.4	48.0	15.7	50.0	16.6	49.2
Std. Deviation:	1.6	2.6	3.8	3.2	2.4	2.8

Standard Practice for Conducting an Interlaboratory Study to Determine the Precision of a Test Method (ASTM E691) Data Analysis[2]

Statistical analysis for each NFRC-accredited testing laboratory of the round robin test data U-factor results were performed. A total of three analyses were performed on the following reported results: U_s, $U_{st[aw]}$, $U_{st[cts]}$. The following table summarizes the analysis of the data:

Table 2a - 1999 results

	Average of Cell Averages	Std. Deviation between Cell Averages	Critical h-statistic
CI_f	19.8	1.626	1.50
CI_g	52.9	2.6537	1.84

Table 2b - 2000 results

	Average of Cell Averages	Std. Deviation between Cell Averages	Critical h-statistic
CI_f	19.9	3.802	2.16
CI_g	53.2	3.214	2.08

Table 2c - Combined 1999/2000 results

	Average of Cell Averages	Std. Deviation between Cell Averages	Critical h-statistic
CI_f	19.9	2.375	2.19
CI_g	53.0	2.748	2.08

Precision and Bias Statement

The 1999 and 2000 interlaboratory comparisons for this procedure have been conducted by the National Fenestration Rating Council (NFRC). These interlaboratory comparisons include the eight NFRC-accredited laboratories that participated in the round robin, with some laboratories having parallel weather side air flow, and others having perpendicular weather side air flow in the thermal chamber. All of the laboratories were required to test the specimen at specified environmental conditions.

Precision

The precision values are presented as an average (mean value from all participating laboratories) thermal transmittance, U_s or U_{st}, and a 95% coefficient of variation. The type of precision described for this round robin is a "Reproducibility Precision Statement". Reproducibility deals with the variability between single test results obtained in different laboratories, each of which has applied the test method to a given test specimen. The summary of inter-laboratory comparison results is provided in the following table.

Table 3a

Reproducibility

Sr. No	Year	Test Specimen	Number of Labs	CI_f	CI_g
1	2000	Non-thermally broken aluminum HS (OX)	8	53.57%	16.92%

Note: Reproducibility variation in percent (between laboratories)

Sr. No	Year	Test Specimen	Number of Labs	CI_f	CI_g
1	1999	Non-thermally broken aluminum HS (OX)	8	22.95%	14.05%

Note: Reproducibility variation in percent (between laboratories)

Table 3b

Coefficient of Variance

Sr. No	Year	Test Specimen	Number of Labs	CV% CI_f	CV% CI_g
1	2000	Non-thermally broken aluminum HS (OX)	8	19.13%	6.04%

Note: CV%= reproducibility coefficient of variation in percent (between laboratories)

Sr. No	Year	Test Specimen	Number of Labs	CV% CI_f	CV% CI_g
1	1999	Non-thermally broken aluminum HS (OX)	8	8.19%	5.02%

Note: CV%= reproducibility coefficient of variation in percent (between laboratories)

Table 3c

Reproducibility

Sr. No	Year	Test Specimen	Number of Labs	CI_f	CI_g
1	1999/2000	Non-thermally broken aluminum HS (OX)	8	32.66%	11.56%

Note: Reproducibility variation in percent (between laboratories)

Table 3d

Coefficient of Variance

Sr. No	Year	Test Specimen	Number of Labs	CV% CI_f	CV% CI_g

1	1999/2000	Non-thermally broken aluminum HS (OX)	8	11.67%	4.13%

Note: CV%= reproducibility coefficient of variation in percent (between laboratories)

Bias

This method has no bias statement as there is no accepted reference value available for comparison.

Summary

This paper provides an initial evaluation of a proposed condensation rating procedure under development by the National Fenestration Rating Council (NFRC). *NFRC 500* testing of a product for determination of a Condensation Index rating shall only be performed if the product can not be simulated. Therefore, simulation is the primary tool used in NFRC 500 to determine the condensation rating.

The evaluation of condensation and its effects on the interior surfaces of fenestration systems is important to designers, specifiers, architects, consumers, manufacturers, building owners and others. NFRC, through the use of *NFRC 500*, provides a means of determining a condensation rating that evaluates the entire product, and provides a comparable rating.

NFRC shall continue to perform annual test laboratory and simulation laboratory round robins. The resultant simulation and test data shall be thoroughly analyzed and compared. Future papers may be developed for peer review and publication, as NFRC is dedicated to continued research in the thermal performance attributes of fenestration products.

Acknowledgments

NFRC-accredited Testing Laboratories

Architectural Testing, Inc. - Fresno, CA
Architectural Testing, Inc. - New Brighton, MN
Architectural Testing, Inc. - York, PA
E.T.C. Laboratories, Inc. - Rochester, NY
Intertek Testing Services - Middleton, WI
STORK/Twin City Testing - St. Paul, MN
National Certified Testing Laboratories, Inc. - York, PA
Quality Testing, Inc. - Everett, WA

NFRC Accreditation Policy Committee Members

John Mumaw (Chair), Dr. Dragan Curcija, Dr. Al Czanderna (1999), Andre Desjarlais, John McFee, Christian Kohler (2000)

References

[1] NFRC 500: Procedure for Determining Fenestration Product Condensation Index Values (Draft Sept. 2000), 1300 Spring Street, Suite 500, Silver Spring, MD 20910.

[2] ASTM E691-99, "Standard Practice for Conducting Interlaboratory Study to Determine the Precision of a Test Method," *Annual Book of Standards*, Vol. 14.02, ASTM International, West Conshohocken, PA, 1999.

APPENDIX

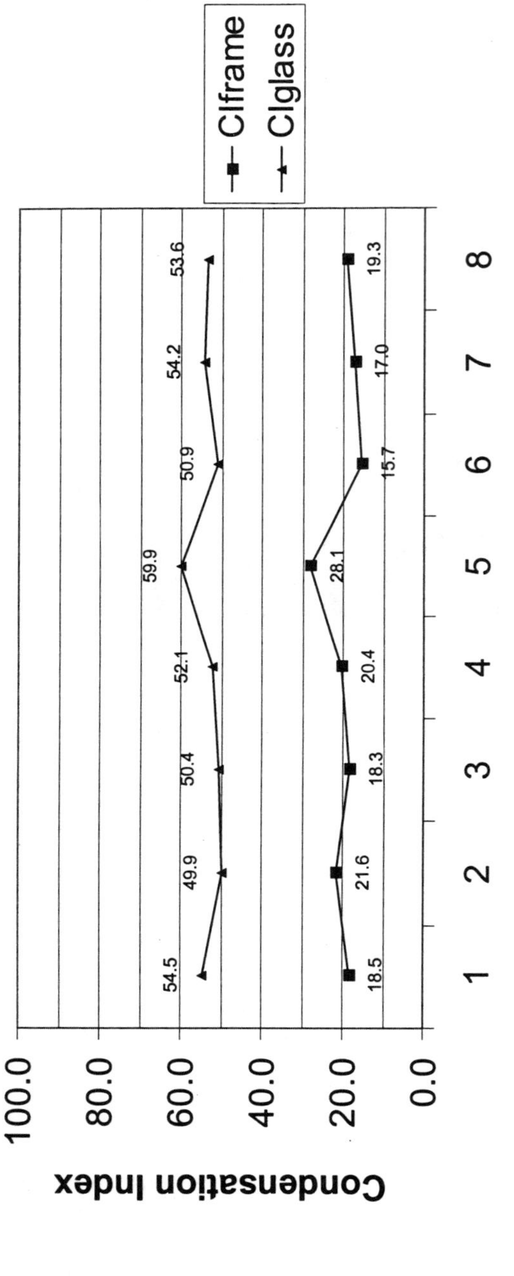

FIGURE 1
Condensation Index Evaluation of
2000 Round Robin Specimen

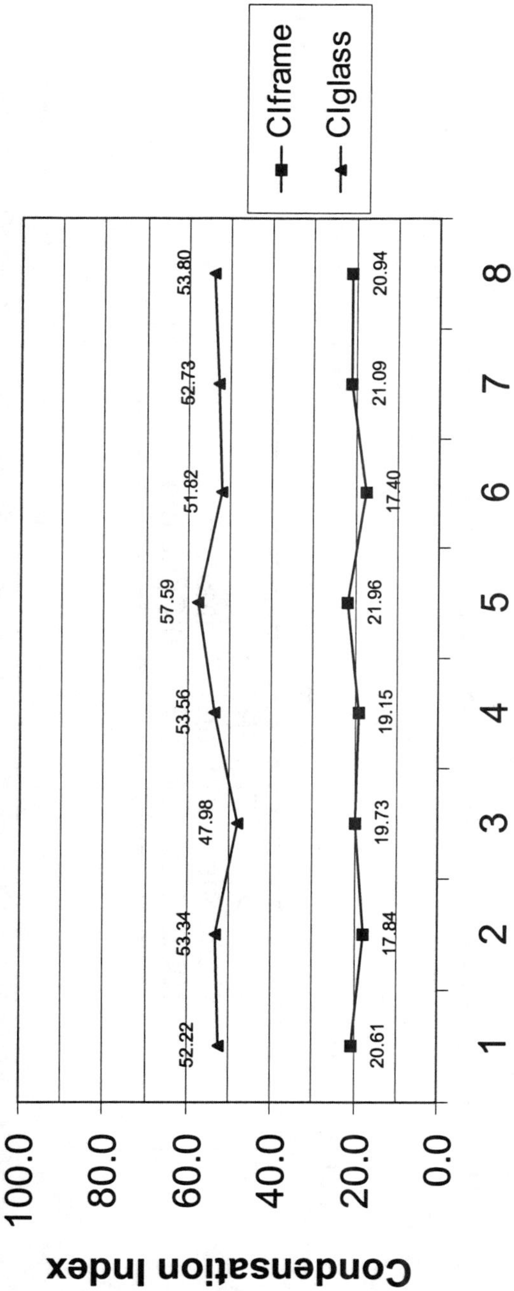

FIGURE 2
Condensation Index Evaluation of
1999 Round Robin Specimen

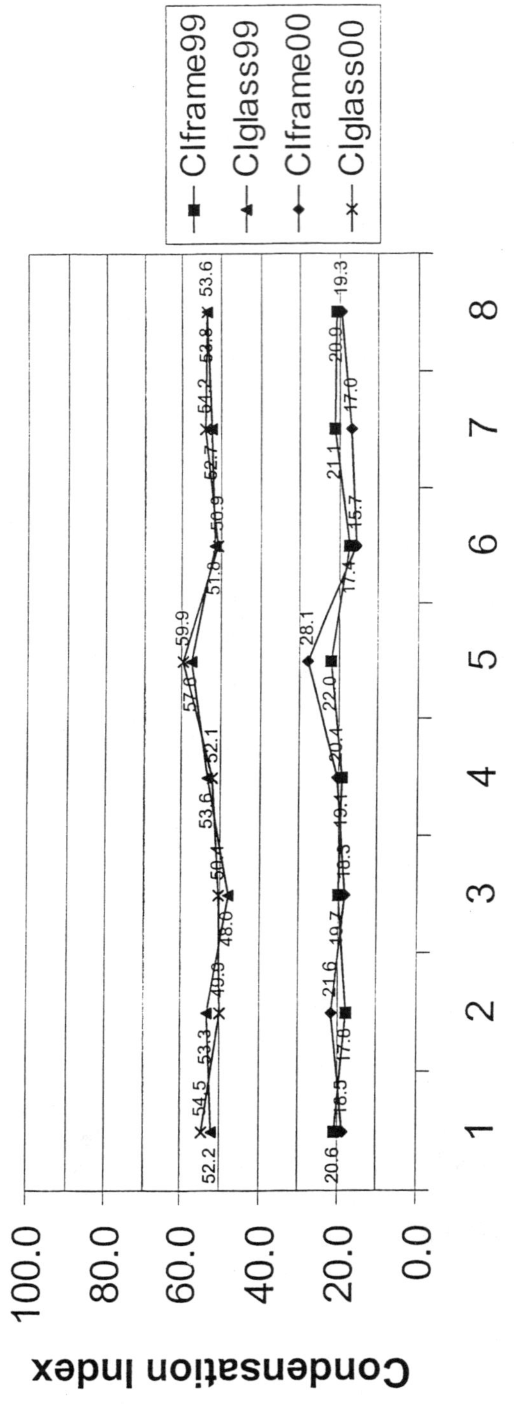

FIGURE 3
Condensation Index Evaluation of
1999/2000 Round Robin Specimen

SECTION IV

Kevin D. Knight,[1] *Bryan J. Boyle,*[2] *and Bert G. Phillips*[3]

A New Protocol for the Inspection and Testing of Building Envelope Air Barrier Systems

Reference: Knight, K. D., Boyle, B. J., and Phillips, B. G., **"A New Protocol for the Inspection and Testing of Building Envelope Air Barrier Systems,"** *Performance of Exterior Building Walls, ASTM STP 1442*, P. G. Johnson, Ed., ASTM International, West Conshohocken, PA, 2003.

Abstract: Premature building failures and increased heating and cooling loads have made building envelope air leakage a growing concern in North America. In Canada, the National Air Barrier Association has made the inspection and testing of air barrier systems a mandatory requirement within its Quality Assurance Program. In the United States, Massachusetts is the first of what is expected to be many states introducing revised Energy Codes making air barriers mandatory in new construction. The Air Barrier Association of America has recently formed and will adopt a similar quality assurance program. Utilizing existing ASTM standards, this paper proposes a protocol for the inspection and testing of air barrier systems to be performed before construction begins, during installation, and post-construction. Studies show that even routine testing of the air barrier system will dramatically improve a building's air permeance.

Keywords: air barrier, air leakage, building envelope, emissions, greenhouse gas, heating and cooling, inspection, testing

[1] President, Retro-Specs Ltd., 395 Berry Street, Suite 8, Winnipeg, MB R3J 1N6.
[2] Marketing Manager, Retro-Specs Ltd., 395 Berry Street, Suite 8, Winnipeg, MB R3J 1N6.
[3] Consulting Engineer, UNIES Ltd., 7666 Dublin Avenue, Winnipeg, MB R3H 0H1.

Introduction

Throughout North America, there is an increasing awareness as to the number of premature building failures that can be attributed to air leakage through the building envelope. The financial, human, and material resources required for these repairs and retrofits, and the costs associated with higher levels of energy consumption and greenhouse gas emissions resulting from heating and cooling loads, are driving the need to adopt new and improved air barrier technologies in order to rectify current conditions.

Air leakage through the building envelope can lead to extensive structural damage. As moisture-laden air moves through the building envelope, the moisture can condense, and under certain environmental conditions, may corrode building components, stain walls and ceilings, stimulate the growth of mold, reduce air quality and damage the exterior façade [1]. A 1993 study estimated that premature deterioration of exterior cladding cost between 159 and 265 million dollars per year in Canada alone [2]. More recent studies have conservatively estimated the cost of building envelope repairs and retrofits in Canada to exceed $500 million per year [3]. While it is difficult to determine to what extent air leakage is responsible for this damage, one study determined that over half of the building set examined had developed building envelope problems within the initial years of occupancy. Most of the problems identified were moisture related, caused either by air leakage or exterior moisture penetration [4].

Stemming from the Kyoto Conference of 1997, countries in North America and abroad are making the reduction of harmful emissions a priority. Increased heating and cooling loads resulting from building envelope air leakage contributes a significant portion of these emissions. To this end, some states are in the process of introducing Energy Code revisions that make the installation of air barrier systems mandatory in new construction.

This paper will identify the functions and requirements of an effective air barrier system, and will discuss why many of today's systems fail to meet these requirements. Visual inspection techniques and quantitative and qualitative testing methods are reviewed, and a protocol for the inspection and testing of air barrier systems is proposed.

Background

Air barriers and other technologies to improve building air permeance were, until recently, utilized primarily in Canadian construction and are only now becoming more prevalent in the United States. Even in Canada, codes and standards were slow to adopt air leakage requirements for buildings. It was not until 1970 that the concept of "airtightness" was officially introduced into the National Building Code of Canada (NBCC), and even at that time, the primary focus was on vapor barriers. A new section – Wind, Water and Vapour Protection – was included in the 1980 NBCC, and while guidelines were presented for vapor barriers, no reference to quantifiable air barrier performance standards was made [5].

In 1995, major revisions to the NBCC took place. Renamed "Separation of Environments", Part 5 of the Code became much more comprehensive and precise in application. This section now applies to condensation control, thermal transfer, and the control of air and moisture through dissimilar environments. In addition, the new Code defined the requirements that successful air barrier systems must possess, and recommended maximum system leakage rates were provided in Appendix A of the Code.

Despite the effect that the revisions would have on both designers, who must ensure that the air barrier material selected meets Code requirements, and on manufacturers, who must provide air permeability specifications on their material, problems still existed in the field. For contractors, responsibility for air barriers remained poorly defined under prevailing trade jurisdictions. Air barrier installers, who were generally untrained for the task, were usually under the direction of the finishing trades, resulting in a lack of continuity in materials and a lack of continuity between building systems.

The National Air Barrier Association (NABA) was officially incorporated in Canada in 1995 with the mission to develop a professional air barrier trade dedicated to installing effective air barrier systems in buildings. NABA has since made the inspection and testing of air barriers a compulsory requirement within its Quality Assurance Program [6]. It is the inclusion of inspection and testing in programs such as this, as well as in documents such as the Canadian National Master Specification, which made the development of a testing protocol a priority.

Development of similar air barrier technology in the United States is now just commencing. While guidelines for building envelope performance, including the control of air movement, had been developed, there was no code in place that would make adhering to these guidelines mandatory practice.

In 1999, United States President Bill Clinton signed Executive Order 13123, establishing new goals for energy management in federal buildings [7]. In addition, many states began the process of revising their Energy Codes, making air barrier systems mandatory. Massachusetts is the front-runner in these developments, introducing an Energy Code making air barriers mandatory in all new buildings. To meet the requirements of the Code, the Air Barrier Association of America (ABAA) was formed in January 2001, incorporating existing Canadian technology, specifications and guidelines in order to expedite the development of a professional air barrier trade in the United States.

Requirements of an Air Barrier System

The aforementioned revisions to Part 5 of the NBCC included four key requirements for successful air barrier systems: airtightness, continuity, structural integrity and durability.

Where a material is to provide the principal resistance to air leakage, the Code states that the air leakage characteristic of the material is not to exceed 0.02 $L/(s/m^2)$ at a pressure differential of 75 Pa. Appendix A of the Code also provides recommended maximum air leakage rates for complete air barrier *systems* in exterior envelopes (Table 1).

The other requirements of the Code state that the air barrier system shall be continuous across construction, control and expansion joints, across junctions between different building assemblies, and around penetrations through the building assembly. Where the air barrier system is subject to wind load, and other elements of separators subject to similar loads, the system must have the ability to transfer the load to the structure. Specifically, the system must be designed and constructed to resist 100% of the specified wind load, and deflections of the air barrier system must not have an adverse affect upon adjoining non-structural elements. Finally, the system must be durable, meaning it must be compatible with adjoining materials and resistant to any mechanisms of deterioration that would be reasonably expected, given the nature, function and exposure of the materials.

Table 1 – *Recommended Maximum Air Leakage Rates.*

Warm side relative humidity at 21°C	Recommended maximum system air leakage rate, L/(s/m^2) at 75 Pa
< 27%	0.15
27 to 55%	0.10
> 55%	0.05

Throughout this paper, for consistency purposes, the key requirements of an effective air barrier system will be as documented in the NBCC. That is to say, the system must meet requirements for airtightness, continuity, structural integrity and durability. The requirements for air barrier systems as specified in the revised Massachusetts State Energy Code are similar to those in the NBCC.

Defining Air Barriers

Based upon these requirements, a material can be defined as an air barrier when it demonstrates an air permeance rating that meets the requirements as prescribed by the NBCC. When installed into a structure, the material must meet the requirements of structural performance and durability, and must have the ability to be joined in a continuous manner throughout the building envelope. An air barrier *system* is the component or combination of components within the building envelope that provide the airtight plane separating areas of differing environments.

In defining air barriers, one must distinguish between air barriers that are "serviceable" or "maintainable," from those which are not. When a chosen material that can meet Code requirements as a system is placed either on the exposed exterior or on the exposed interior to prevent the movement of air through the building envelope, it is said to be a maintainable air barrier. In other words, its position in the wall is such that it can be serviced. When the air barrier is placed mid-wall, where its wall position is such that it cannot be serviced, it is said to be non-maintainable.

The benefit of a maintainable air barrier system is just that – it is maintainable. However, the nature of this type of system makes it difficult to achieve continuity between the individual components that comprise the system. It is much easier to achieve continuity when installing mid-wall air barriers such as bituminous membranes, spray-applied polyurethane or rubber polymer, which provide a more durable system [8]. However, as this form of system is non-maintainable, it is imperative that *prescriptive* testing and inspection be performed to ensure that the installation is correct and the air barrier is functional prior to cladding being installed over the completed system.

Why Things Go Wrong

The combination of system design, materials and workmanship will determine whether or not the completed air barrier system will perform to specification. Inadequacies in any of these elements will have a negative impact on the performance of the system in the short and/or long run.

The objective of the air barrier system on a general level is adherence to the requirements outlined in the NBCC, or on a specific level, to the specifications particular to that project.[4] However, installing the system to specification does not guarantee that the system will operate effectively, as there may be instances where design deficiencies lead to failure in the system. Lack of continuity between components that make up the system, or an incompatibility between the materials that make up these components, are commonly occurring problems. Continuity may also be jeopardized when the details themselves are difficult or even impossible to construct effectively.

There is a wide range of commercially available air barrier membrane materials that exceed the materials requirements outlined in the NBCC [9]. When selecting other materials however, care must be taken to ensure that the on-site fabrication and/or installation of the material does not compromise its air permeance to the extent that the air barrier no longer satisfies Code requirements. For instance, spray-on air barrier as a material may exceed Code requirements, but as the material is mixed on site, incorrect fabrication may render it ineffective. As well, certain materials used as an air barrier within the system may react adversely to bonding agents required for installation, for example certain primers, or even to adjacent materials themselves, such as caulking.

The final quality of the air barrier system will ultimately depend upon the quality of installation. It would be fair to say that in most areas of North America, the quality of installation is questionable. As this is an emerging technology, it is a new and unfamiliar task for trades who install air barriers and there is a distinct lack of skilled labor in this regard. The result is that trades that are responsible for the airtightness of the building may have little or no knowledge of the requirements of the completed system.

Several problems arise from the fact that currently the installation of the air barrier is a shared responsibility between trades installing the individual components of the system. While specific trades may be responsible for a "section", such as window-to-wall or wall-to-roof, no one trade is responsible for ensuring that the individual sections are joined in an effective manner. As trade jurisdictions overlap, there is frequently a lack of communication and cooperation between the trades, leading to scheduling conflicts. In many instances, air barrier materials are being installed only shortly before the finishing components, resulting in the air barrier being covered before the system is complete. Also, the system may be compromised if the finishing veneer is fastened through the air barrier.

While the NBCC recommends system air leakage rates, they are merely recommendations. In the past, the lack of definitive performance standards and means by which to measure air leakage rates on a particular installation resulted in systems that

---

[4] The air barrier system must also adhere to any state/provincial or municipal building codes that may override the NBCC or other national code. For the purposes of this paper, when discussing adherence to the NBCC, this will assume adherence to these other codes.

failed to meet NBCC recommendations. While the lack of performance standards remains a concern, the problem is being addressed on numerous levels. At the government level, improvements are being made to existing codes and standards, taking advantage of more modern inspection and testing technologies. At the industry level, quality assurance programs are being developed and implemented, with education provided to both the designers and the workforce. Finally, manufacturers are testing their materials more stringently, both as materials and in systems.

The Protocol of Inspection and Testing of Air Barrier Systems

In developing a regimen for the inspection and testing of air barrier systems, several factors were considered; what are the testing criteria, what is being tested, when and how often should inspection and testing be conducted, who should perform the tests, and what methods should be used. Obviously, the air barrier should meet the NBCC requirements, all prevailing codes and/or specifications (for instance, the National Master Specification, or State Energy Codes), and any other specifications particular to that project. In addition, the performance of both materials and the system should be evaluated under the guidelines of the Canadian Construction Materials Centre (CCMC) or similar body. The CCMC has recently prepared technical guides that outline test criteria to determine structural integrity, air leakage and durability of the material or system.

Due to the varying nature of air barrier systems and materials, any generic inspection and testing protocol must be comprehensive enough to encompass any type of system. To illustrate this point, assume a testing regimen called only for post-construction testing. For a system installed mid-wall, there are limited means to test for air leakage once the building is complete. Even if testing were to indicate that the system failed, not only would it be difficult to determine which component in the system failed, but it may be expensive to repair a system that has already been covered with finishing materials. It has been estimated that the cost to repair a failed air barrier could be greater than 50 times the cost of installing it correctly the first time [10]. It should then be obvious from this example that inspection and testing should be performed in three stages: pre-construction, construction-in-progress and post-construction.

Note that this represents an "optimal" degree of inspection and testing. In reality, not all projects will specify all three phases of testing all of the time. Specific job testing may include one or more of the phases, but not necessarily all three. Depending upon the performance requirements of the building, or the degree of complexity of the building envelope system, pre- or post-construction inspection and testing may not be required. In this sense, the protocol is system specific, and it will be up to both the designer and the building owner to determine the degree to which inspection and testing will be performed. However, some inspection and testing during the construction process should be mandatory on all projects.

Several ASTM standards will be utilized in the proposed inspection and testing protocol (Table 2). While there were ASTM recognized test methods that could be utilized for pre- and post-construction air leakage testing, prescriptive testing during construction posed some difficulties. The methods included in ASTM Practices for Air Leakage Site Detection in Building Envelopes and Air Retarder Systems (E 1186) were, in most cases, non-applicable for work in progress due to reliability or on-site practicality. However, recent revisions to E 1186 have made it practical to test for air leakage during construction. Not only have test methods been included that can be performed inexpensively and without disrupting the critical path of construction, in many cases the tests can be performed by the installers themselves. With both self-testing and third party testing, there is an increased likelihood of achieving an effective air barrier system on the first installation.

TABLE 2 – *Standards Utilized in Proposed Inspection and Testing Protocol.*

Standard	Title
ASTM D 4541	Test Method for Pull-Off Strength of Coatings Using Portable Adhesion Testers
ASTM E 283	Test Method for Determining the Rate of Air Leakage Through Exterior Windows, Curtain Walls, and Doors Under Specified Pressure Differences Across the Specimen
ASTM E 330	Test Method for Structural Performance of Exterior Windows, Curtain Walls, and Doors by Uniform Static Air Pressure Difference
ASTM E 741	Test Method for Determining Air Change in a Single Zone by Means of a Tracer Gas Dilution
ASTM E 779	Test Method for Determining Air Leakage Rate by Fan Pressurization
ASTM E 783	Test Method for Field Measurement of Air Leakage Through Installed Exterior Windows and Doors
ASTM E 1186	Practices for Air Leakage Site Detection in Building Envelopes and Air Retarder Systems • 4.2.1, Building Depressurization (or Pressurization) with Infrared Scanning Techniques • 4.2.2, Smoke Tracer in Conjunction with Building Pressurization or Depressurization • 4.2.3, Building Depressurization (or Pressurization) in Conjunction with Airflow Measurement Devices, or Anemometers • 4.2.4, Generated Sound in Conjunction with Sound Detection • 4.2.5, Tracer Gas • 4.2.6, Chamber Pressurization or Depressurization in Conjunction with Smoke Tracers • 4.2.7, Chamber Depressurization in Conjunction with Leak Detection Liquid
ASTM E 2099	Practice for the Specification and Evaluation of Pre-Construction Laboratory Mockups of Exterior Wall Systems
CAN/CGSB-149.10-M86	Determination of the Airtightness of Building Envelopes by the Fan Depressurization Method
CAN/CGSB-149.15-96	Determination of the Overall Envelope Airtightness of Buildings by the Fan Pressurization Method Using the Building's Air Handling System

It is important to distinguish between inspection and testing. Inspection is simply a visual examination of a specific detail, specific components, or the system. Testing involves taking some form of measurement, be it qualitative or quantitative, to determine whether the component(s) or system in question has the ability to perform a given function, and in some cases, to what degree the function can be performed. For the purposes of this paper, unless otherwise specified, when general reference is made to "testing", it includes "testing and inspection".

Pre-Construction Phase

The designer must ensure that the system as designed will function as per NBCC requirements. Prior to the submittal of contract drawings and bid documents, project plans and specifications should be reviewed by the designer and third party building envelope consultant to confirm that there are no problems inherent in the design. The review should consist of an examination of individual details, components, materials and material compatibility, and how all of these work within the confines of the system.

Before construction begins, an orientation meeting should be called by the third party consultant (and arranged by the general contractor) to discuss various aspects of the project. Present at the meeting should be the owner, designer, consultant, general contractor and sub-contractors. At the meeting, all parties will be given a chance to view the plans, specifications and shop drawings, and to voice any concerns they may have in this regard. Other items on the agenda should include compatibility of materials, and construction sequencing.

Once the design aspects have been considered, a mockup of the system should be constructed and tested. ASTM Practice for the Specification and Evaluation of Pre-Construction Laboratory Mockups of Exterior Wall Systems (E 2099) describes construction and documentation procedures and protocol to assist in the specification and evaluation of pre-construction laboratory mockups.

In evaluating the ability of an air barrier system to perform as required, it is important to consider components or sections of the system where failures most often occur. Weaknesses in an air barrier system commonly occur at window surrounds, fastener penetrations, and junctions between dissimilar materials and building components. While in some cases it may be impractical to construct a mockup to test every detail of the system, the mockup must be representative of what will be constructed, including the various wall conditions that will be encountered. The designer will specify what details should be included in the mockup. For some systems, several mockups may be required to include all of the details specified.

To determine the air leakage rate of the system, quantitative testing should be performed on the mockup in accordance with ASTM Test Method for Determining the Rate of Air Leakage Through Exterior Windows, Curtain Walls, and Doors Under Specified Pressure Differences Across the Specimen (E 283). This test is performed under laboratory conditions, and is intended to measure only leakage associated with the assembly, and not the installation. In some cases, the mockup is constructed on site, to be installed as part of the actual system, subject to design specifications. ASTM Test Method for Field Measurement of Air Leakage Through Installed Exterior Windows and Doors (E 783) should be used under these circumstances. Some designers may even require that mockups be constructed both for testing under laboratory and site conditions. In all instances, testing should be conducted by a third party.

Both of these test methods provide a quantitative measurement of the air leakage rate of the system. In addition, by exposing the mock-up to specified pressure differentials, it can be determined whether the structural stability of the system and the strength of the bond between the membrane and substrate are sufficient to withstand the loads likely to be placed against it. ASTM Test Method for Structural Performance of Exterior Windows, Curtain Walls, and Doors by Uniform Static Air Pressure Difference (E 330) can be performed in this regard.

While these procedures can be used to evaluate the design aspects of the system, they do not take into account variables such as site conditions, scheduling or workforce skill level. As well, since specifications usually allow for a choice of materials, the material installed on site may differ from the originally tendered material. Hence, the system that was tested may differ from the system as it is to be installed. It is imperative, then, that testing be performed on the actual system as it is installed and as it is being installed, in order to best determine whether the system satisfies NBCC requirements.

Construction-in-Progress Phase

This phase represents the most comprehensive degree of testing in the regimen, differing from pre- and post-construction testing in that it allows for the testing of individual components that make up the completed system. The rationale behind testing during construction should be obvious – it is both easier and less costly to correct a problem if it is discovered early in construction than once the building has already been completed. Testing during this phase will be conducted both by a third party and the contractor installing the air barrier.

Workforce Testing – With the introduction of professional air barrier trade associations in both Canada and the United States offering education, training and certification of the workforce, every effort should be made to ensure that the installers are licensed by such an organization. Organizations such as NABA and ABAA utilize quality assurance programs that require the workforce to not only test their installations, but to provide written confirmation that testing was performed, and that the detail tested met requirements.

A certified installer should perform visual inspections daily. Before any installation occurs, the installer should examine the material, ensuring that it is compatible with the bonding agent and adjoining materials, and if fabricated on site, that it was done so according to manufacturers' recommendations. The substrate should also be inspected to ensure it is dry, clean and primed (according to manufacturers' specifications) before any material is applied. The temperature of the substrate should also be measured.

The installer should be aware of the construction schedule to ensure that different components can be joined in a continuous manner. As well, he must consider the "in-progress" building conditions, and where the exposed air barrier is in relation to environmental conditions. For example, modified bituminous membranes that have been left exposed to direct sunlight for extended periods may wrinkle and leak at the seams.

With the aforementioned revisions to E 1186, installers are able to test various details for air leakage. Using E 1186 4.2.7, seams and penetrations can be qualitatively tested for airtightness. In order to establish whether any major problems exist with either the materials or the workforce, a greater percentage of self-testing should be concentrated toward the front end of the project. For instance, if testing revealed that a high percentage of fasteners leaked, there may be a problem with the particular type of fastener as it has been used on this project. Discovering the problem at an early stage makes it much easier to rectify.

For the remainder of the project, unless the specifications call for complete airtightness (as would be required for a high-performance building such as a disease control center, swimming pool or art gallery), a reduced sample area representative of the system is appropriate. The presence of some factors, such as extreme weather conditions or a changing of the workforce or material will influence the size of the selected sample area.

The strength of the bond between the membrane and substrate should be tested in accordance with ASTM Test Method for Pull-Off Strength of Coatings Using Portable Adhesion Testers (D 4541). This test method provides a quantifiable measure of the adhesion of a coating by determining either the greatest perpendicular force that a surface area can bear before a plug of material is detached, or a qualitative result if the surface remains intact at a prescribed force. The test is used to determine whether the strength of adhesion between the membrane and substrate is in compliance with manufacturer's recommendations. Testing should be performed regularly and every time conditions change, for example, changes in environmental conditions, substrate conditions, material batch, and workforce.

Third-Party Testing - The goal of third-party testing at this phase is two-fold: to provide quality assurance testing of the installation as it progresses and to provide a benchmark, both visual and quantifiable, for the installers as to the degree of air leakage that is acceptable.

Visual inspections should be conducted prior to any testing. If obvious deficiencies are present, for example flutes or areas of unbonded or discontinuous membrane, testing should not proceed until the deficiencies have been addressed. The materials as fabricated or installed should be inspected to ensure they are in keeping with manufacturers' or designers' specifications. Once it has been established that deficiencies cannot be visually identified and that the sample area is in keeping with the above conditions, testing may proceed.

Once again, testing should be performed in accordance with E 783 if a numerical benchmark for air leakage allowance is needed. The area or detail tested under E 783 should also be tested in keeping with E 1186 4.2.6 in order to provide installers with a visual benchmark.

As air barriers are still only an emerging industry in both Canada and the United States, many stakeholders have yet to fully understand the importance of an effective air barrier system, and henceforth, the importance of a complete testing regimen. Unfortunately, the degree to which third party testing will occur depends not only upon the performance criteria of the building and the environmental conditions where the building is situated, but also concurrently upon the designers and project budget. Ideally, the regimen would be performed as follows: as per the pre-construction phase, there would be a comprehensive review of drawings and details. The system would then be tested in accordance with E 283 in the laboratory before the start of the project. Once construction has begun, but prior to the air barrier being installed, a mock-up on site should be tested in accordance with E 783. Now, as the air barrier is being installed, visual inspection should be performed. The installation should again be tested in keeping with E 783, E 1186 and D 4541 to provide the benchmarks as well as to test the system as it is installed by the labor force.

In most current situations, where budgetary restraints are placed upon testing, certain steps may be omitted (for example, the E 783 may only be performed once). The designer will determine the optimal inspection and testing regimen based upon the budgetary constraints placed upon the project by the building owner.

Similar to the trades, third party testing agents can test specific details in accordance with E 1186 and D 4541. Once again, the amount of testing will be dependent upon the aforementioned factors.

Post-Construction Phase

Once the building is complete, there are a variety of ASTM and Canadian General Standards Board (CGSB) tests that can be used to determine the airtightness of the building. A report released by the Canadian Mortgage and Housing Corporation (CMHC) has outlined several options for testing completed air barrier systems [11]. Quantified testing can be performed in accordance with ASTM Test Method for Determining Air Change in a Single Zone by Means of a Tracer Gas Dilution (E 741) and ASTM Test Method for Determining Air Leakage Rate by Fan Pressurization (E 779).

E 1186 provides several acceptable and practical test procedures: sections 4.2.1 and 4.2.2 both provide reliable results. CAN/CGSB Determination of the Airtightness of Building Envelopes by the Fan Depressurization Method (CAN/CGSB-149.10-M86) or CAN/CGSB Determination of the Overall Envelope Airtightness of Buildings by the Fan Pressurization Method Using the Building's Air Handling System (CAN/CGSB-149.15-96) may also be used.

While the tests do provide reliable results as to the airtightness of the building, there are some concerns. The tests only identify the presence of air leakage; they do not identify the locations of the leak. Also, the numerical results generated may indicate that the building meets airtightness requirements, despite the fact that one or more key components in the system have failed. So while the tests are helpful in determining whether the system has been constructed effectively, it does emphasize the need for testing both before and during construction. Note also that these tests may require extensive preparation and the presence of an engineer to generate the numerical results.

Does Testing Improve the System's Performance?

An internal study conducted by the authors set out to estimate the monetary and environmental impacts of improving the airtightness of building envelopes [*12*]. The study determined the effect that testing had upon the airtightness of the building envelope, and from there estimated the reduction in heating load requirements and greenhouse gas emissions, and the savings related to increased building envelope life cycles. The study was conducted in 1997 using both laboratory and field results.[5]

Results from the study generated information as to the impact of routine airtightness testing of specific construction details. Specifically, the study found that when air barriers are tested, the number of leaks at the detail decrease, the average size of the leaks at details decrease, and that routine testing of a specific detail identifies materials and/or construction practices that negatively affect airtightness levels at the detail. The result is an overall reduction in air leakage for the air barrier system, the amount of which is dependant upon the reduction in the actual number of details that leak and the reduction in average leakage area at the details which still leak. The study used masonry ties as an example of a construction detail prone to leakage (Table 3). The potential reductions in the leakage area of the masonry ties were estimated to be between 77 and 91% resulting from routine air leakage testing, and it could be reasonably estimated that for most other details, similar results would be found.

TABLE 3 — *Estimates of Potential Reduction in Leakage Area.*

	Typical			Achievable Goal		
	Percent leaking	ELA per tie (cm^2/tie)	Overall ELA (cm^2/m^2)	Percent leaking	ELA per tie (cm^2/leak)	Overall ELA (cm^2/m^2)
Cast-in-place ties, masonry	75	0.040	0.168	30	0.009	0.038
Cast-in-place ties, drywall	90	0.089	0.374	40	0.008	0.034
Through-wall ties, masonry	25	0.0015	0.0063	3	0.0002	0.0008

The study concluded that higher levels of airtightness in buildings would in turn result in a lower life cycle cost for building envelopes by decreasing the heating and maintenance costs for building owners. By achieving the least stringent NBCC recommendation for system airtightness (0.15 L/s/m^2 at 75 Pa), there would be an estimated savings of $5.9 million per year in heating energy requirements for the building set. This translated to a savings of between 1.91 and 5.96 \$/m^2 of wall area per year for electrically heated buildings, and between 0.63 and 1.99 \$/m^2 of wall area per year for natural gas heated buildings.[6]

The study also found that air leakage related deterioration could have a significant impact on the achieved life of the building envelope. It was estimated that premature deterioration of building envelopes result in an increase in the life cycle cost of the building envelope between $29 and $109 per square meter of wall area in present worth. For the building set considered in the study, the savings in future cladding replacement costs resulting from air leakage was estimated to be $3.0 million per year.

---

[5] The building set consisted of 3,563 buildings in Winnipeg, Manitoba, and was considered representative of the buildings in Winnipeg, which could utilize air barrier membrane technology.

[6] Based upon 1997 energy prices. Today, at Manitoba rates, savings would be between 0.95 and 3.00 \$/m^2 of wall area per year for natural gas heated buildings. The savings for electrically heated buildings would not change.

By reducing both heating requirements and cladding replacements, there was a corresponding reduction in greenhouse gas emissions, landfill use, and peak electric demand. It was estimated that the annual reduction in carbon dioxide emissions due to reduced heating consumption and reduced building envelope failure was 62,300 tons per year and 600 tons per year respectively. Peak electrical demand reduction was estimated to be 26 MW and the reduction in cladding materials deposited in landfill was estimated to be 2100 tons per year.

Conclusion

With the high number of premature building envelope failures, and a global concern regarding greenhouse gas emission levels, it is imperative that air leakage be minimized in commercial buildings. The introduction of more stringent codes and specifications, increased air barrier materials testing, and the development of industry quality assurance programs indicate that the problem of insufficient air leakage control is being addressed by the various industry stakeholders. There is an obvious need for a protocol for the inspection and testing of air barrier systems to ensure that the systems are performing as expected and required.

If air barriers are being installed into buildings to reduce air leakage, it is imperative that they be installed correctly. By utilizing existing ASTM standards in the manner outlined in this paper, building air leakage rates can be improved to the point that they meet the recommendations as outlined in Appendix A of the National Building Code of Canada and emerging state energy codes.

While inspection and testing will improve the performance of the installed system, it is not the sole solution to the problem. Owners have to be educated to the impact of building envelope air leakage. Designers should be informed about available quality assurance programs, utilizing ASTM standards, which address air leakage. Educational programs must be in place in order to develop a skilled air barrier trade. With all of these forces working in unison, we can begin to reap the benefits associated with improved airtightness in buildings.

References

[1] Dalgleish, R. and Knight, K. D., "Air Barriers: Building Essentials," *Construction Canada*, Vol. 41, No. 6, 1999, pp. 10–14.

[2] Hanscomb Consultants Inc., "National Roofing and Exterior Cladding Market Assessment Study," National Research Council Canada, Ottawa, ON, November 1993.

[3] Rousseau, J., Canadian Mortgage and Housing Corporation, Ottawa, ON, letter to building envelope council members, June 1995.

[4] Drysdale, R. G., "Construction Problems in Multi-Family Residential Buildings," Canada Mortgage and Housing Corporation, Ottawa, ON, March 1991.

[5] Knight, K. and Samuda, M., "Newly Developed Means of Testing Air Barriers During Construction: The Method and the Industry," *Fourth Energy-Efficient New Construction Conference*, Vancouver, BC, 1996, pp. 131–140.

[6] Building Professionals Consortium, "Professional Contractor Quality Assurance Program," National Air Barrier Association Inc. and Building Professionals Consortium, Winnipeg, MB, April 1997.

[7] Tsai, A., "Berkeley Lab Acts to Implement Presidential "Greening" Order," *EETD News*, Winter 2000, pp. 5–8.

[8] Persily, A. K., "Envelope Design Guidelines for Federal Office Buildings: Thermal Integrity and Airtightness," U.S. Department of Commerce, Gaithersburg, MD, March 1993.

[9] AIR-INS Inc., "Air Permeance of Building Materials," Canada Mortgage and Housing Corporation, Ottawa, ON, June 1988.

[10] Kirbyson, G., "U.S. Deals in the Air For City Energy Firm," *Winnipeg Free Press*, January 8, 2001, p. B4.

[11] Morrison Hershfield Ltd., "Commissioning and Monitoring the Building Envelope for Air Leakage," Canada Mortgage and Housing Corporation, Ottawa, ON, November 1993.

[12] Sharp, J. F., Phillips, B. G. and Knight, K. D., "Air Barrier and Building Envelope Environmental Impact Study," Manitoba Energy and Mines, Winnipeg, MB, April 1997.

Heinz R. Trechsel [1]

Overview of ASTM MNL 40, Moisture Analysis and Condensation Control in Building Envelopes

Reference: Trechsel, H. R., **"ASTM MNL 40, Moisture Analysis and Condensation Control in Building Envelopes,"** *Performance of Exterior Building Walls, ASTM STP 1422,* P. G. Johnson, Ed., ASTM International, West Conshohocken, PA, 2003.

Abstract: Following up on ASTM MNL 18, *Moisture Control in Buildings*, ASTM Committees C16 and E06 now co-sponsor ASTM MNL 40, *Moisture Analysis in Buildings*. MNL 40 expands the chapters on Modeling and on Design Tools and adds to the chapters on Weather Data and Materials Data of MNL 18, published in 1994.
The purpose of the new manual is to provide an overview of the current state of moisture analysis in buildings, to encourage the use of moisture analysis by building designers and practitioners, and to provide the necessary technical background for conducting moisture analysis as an integral part of building design. The manual consists of ten chapters written by individual experts. Included are an extensive Glossary of terms used in moisture analysis and a moisture primer addressing the reader with no or with only limited understanding of hygrothermal mechanisms. Other chapters discuss weather and material data and failure criteria. The manual continues with two overviews of analysis methods, a chapter on manual methods and chapters on the models MOIST and WUFI ORNL/IBP. The manual concludes with a Look to the Future. Enclosed with the manual are CD-ROM disks with the programs for MOIST and WUFI ORNL/IBP, including the necessary owners manuals, and with two conversion programs for properties of air.

Keywords: buildings, moisture analysis, building envelopes, condensation, modeling, design tools, mold, mildew

Introduction

If we compare the practice of the design of buildings for moisture resistance performance with the practices for structural performance, we find that structural adequacy is determined by rigorous analysis. Structural properties of materials, shape of components, type of connections, loading conditions, etc. are all considered and their numerical values incorporated into the calculations. On the other hand, the design for moisture resistance is still essentially relying on simple prescriptive rules, most of which were established over fifty years ago. Although structural design also starts with some

[1] H. R. Trechsel Associates, Arlington, VA, and Engineering Field Activity Chesapeake, Naval Facilities Engineering Command, Washington, DC. The opinions expressed herein are those of the author and do not necessarily reflect those of any Government agency.

similar prescriptive rules—such as span to depth ratios for beams—the final design is verified by numerical analysis. In moisture design, we seldom, if ever, verify the adequacy by analysis, but are satisfied with initial assumptions based on prescriptive rules. One reason for this discrepancy may be that moisture does not normally cause catastrophic and life threatening failures, while structural failures frequently do so. Another reason is that the analytical tools were either missing or cumbersome to use, and provided only snapshots for a particular set of conditions. A third reason is that moisture-related properties of materials, unlike structural properties of materials, were largely unknown or poorly understood. It is the purpose of MNL 40, *Moisture Analysis for Buildings*, to provide both the practitioner and the building researcher with an understanding of the tools available and to introduce moisture analysis and moisture engineering into the design practice of buildings.

The traditional remedy to solve moisture problems in buildings were prescriptive rules: In cold climates install a vapor retarder on the warm side of the insulation in building walls and roofs, that is on the interior side. In warm climates, place the vapor retarders also on the warm, in this case exterior, side of the insulation. The rules were less clear regarding what were cold and warm climates. In addition, there exist separate recommendations in various books, symposium proceedings, and in stand-alone papers which discuss concepts and suggestions for preventing moisture damage in building envelopes.

Most of these suggestions and recommendations are readily accessible and useful to the building designer/practitioner. In general, only the most basic suggestions and recommendations are currently referenced in building regulations and in Guide Specifications. In addition, many of these suggestions discuss items largely outside the control of the designer, such as on-site workmanship and on-site quality control.

The practice of moisture analysis can expand on the prescriptive rules, and can provide the designer and the building owner a measure of assurance that major moisture damage due to condensation will not occur. However, moisture analysis cannot substitute for proper on-site workmanship and quality control, and cannot rectify poor detailing of joints and lack of proper flashing.

Recognizing the need for bringing moisture analysis to the attention of a broad audience of building practitioners and building researchers, the Building Environment and Thermal Envelope Council (BETEC) [1] conducted in 1996 a Symposium on Moisture Engineering. The symposium presented an overview of the current state-of-the-art of moisture analysis. The consensus of the participants was that Moisture Analysis was now practical as a design tool, and that it should be given preference over the simple application of the prescriptive rules. However, the consensus also was that the Architect/Engineer community was not ready to fully embrace the analytical approach. In response to this consensus, BETEC, supported by the Department of Energy (DOE), the National Institute of Standards and Technology (NIST), and the National Institute of Building Sciences (NIBS) prepared and conducted a two-day pilot tutorial primarily for practicing architects and engineers. The tutorial was conducted in September 1997 and is described in a paper presented at a ASTM Symposium in the spring of 1998 [2]. Based on the experience from this tutorial, the development of ASTM Manual MNL 40 was co-sponsored by ASTM Committees C16 and E06.

MNL 40 expands and elaborates on the relevant chapters of MNL 18, *Moisture Control in Buildings* [3]. Specifically, MNL 40 builds on MNL 18 Chapter 2, "Modeling Heat, Air, and Moisture Transport through Building Materials and Components;" on Chapter 3, "Relevant Moisture Properties of Building Construction Materials;" Chapter 7, "Climate;" and on Chapter 11, "Design Tools." MNL 40 also includes a Chapter on Moisture Primer, which is intended to provide the beginner an understanding of the physics of moisture movement, condensation, and evaporation, but the manual is not intended as a textbook for Hygrothermal Dynamics or advanced building physics. For a deeper understanding of these subjects, the reader is referred to existing textbooks and handbooks, such as the ASTM Handbook of Fundamentals and to ASTM MNL 18.

Prescriptive Rules and their Origin

In 1948, the U.S. Housing and Home Finance Agency (a forerunner of the current Federal Housing Administration) held a meeting attended by representatives of building research organizations, homebuilders, trade associations, and mortgage finance experts on the issue of Condensation Control in Dwelling Construction [*4*]. The meeting focussed on vapor diffusion in one-and two-family frame dwellings in cold weather climates. The meeting resulted in the establishment of the rule to place a vapor barrier (now called a vapor retarder) on the warm side of the thermal insulation in cold climates and established that a vapor barrier (retarder) shall be a membrane or coating with a water vapor permeance of one Perm or less. One Perm is 1 gr/hr·ft^2·in.Hg (57 ng/s·m^2·Pa). The rule was promulgated through the FHA Minimum Property Standards [*5*]. It still is referenced unchanged in some construction documents.

The above rule was based on the assumption that diffusion through envelope materials and systems is the governing mechanism of moisture transport leading to condensation and eventual degradation of the building envelope. Since 1948, and particularly since about 1975, research conducted in this country and abroad has shown that infiltration of humid air into building wall cavities and the leakage of rainwater are significant, in many, if not most, cases governing mechanisms of moisture transport. Accordingly, the original rule was expanded to include the requirements for restricting air infiltration by installing air barriers and for flashing to exclude rainwater. Attempts were also made to cover climates other than cold, and to include building and construction types other than frame buildings.

The current expanded rules have greatly increased the validity and usefulness of the Prescriptive Rules. However, the rules still do not fully recognize the complexities of the movement of moisture and heat in building envelopes. Specifically:

- The emphasis on either including or deleting a separate vapor retarder ignores the effect of the hygrothermal properties of other envelope materials on the flow of moisture through the building envelope.
- Climate as the only determining factor is inadequate to determine whether a vapor retarder should or should not be installed. Indoor relative humidity and the moisture-related properties of all envelope layers must also be considered.
- The two climate categories "cold" and "warm" have never been adequately or consistently defined, and large areas of the contiguous United States do not fall under either cold or warm climates, however defined. For example:

For All Climates:

- ASHRAE [*6*], in 1993, presented Condensation Zones based on design temperatures.

For Cold Climates:

- Lstiburek [*7*] suggests 4,000 Heating Degree Days or more.
- The U.S. Department of Agriculture [*8*] recommends an average January temperature 35°F or less.

For Warm Climates:

- ASHRAE [*9*] established criteria based on the number of hours that the wet bulb temperature exceeds certain levels.
- Odom [*10*] suggests Average monthly latent load greater than average monthly sensible load for any month during the cooling season.
- Lstiburek [*11*] suggests defining hot-humid region as one receiving more than 20 in. (500 mm) of annual precipitation and having the monthly average outdoor temperature remaining above 45°F (7°C) throughout the year.

It is apparent that prescriptive rules alone will not assure that building envelopes are free of moisture problems. However, today we have a great number of moisture analysis methods, both manual and computer based models that allow the analysis of building envelopes to determine moisture content, surface temperatures, and condensation planes. The manual methods use predetermined conditions; their results are dependent on the validity of the conditions selected, and they do not provide a measure of the duration of moisture excursions. The computer models are based on weather data, most using hourly readings from a set of an "average" year for a specific location and do provide the duration of likely moisture excursions.

Analytical Methods and Models and Their Limitations

The progress made in the development of computer-based analysis methods, or models since the publication of MNL 18 in 1994 has been spectacular. At last count, there exist now well over 30 models that analyze the performance of building envelopes based on historical weather data. The models vary from simplified models useable by building practitioners on Personal Computers to sophisticated models which require specially trained experts and which run only on mainframe computers.

The simpler models may or may not include the effect of moisture intrusion due to air and water infiltration. The more sophisticated models are excellent tools for building researchers and, as a rule, do include the effects of rainwater leakage and of air infiltration. Although air infiltration and water leakage can be significant causes of moisture distress in building envelopes, analytical models that do not include these factors are still useful to the designer, because:

- The input data for air infiltration and water leakage for a specific design is often not available or is unreliable. Also, much of the performance of joints depends on field workmanship and quality control over which the designer seldom has significant control and which are largely unknown at the design stage.
- Air infiltration and rainwater leakage, unlike diffusion, occur at distinct leakage sites. These are seldom evenly distributed over the building envelope. The effect of air and water leaks are bound to be localized, their locations unknown at the design stage, their exact locations frequently difficult to ascertain even in existing buildings.
- Both air and water leaks are transient in nature, with durations measured in hours, days, or weeks. Rainwater leakage and air infiltration depend on wind direction. Rain falls one day, leaked water can dry during the next day or week. Moist air moves into the envelope one hour, one day; and the next hour, next day dry air may enter the envelope, wetting turns to drying. By contrast, diffusion mechanisms operate generally on a longer time horizon, frequently for weeks, months, or over an entire season before they change from inward to outward, or vice versa.

The use of Moisture Analysis alone does not guarantee moisture resistant buildings. Careful detailing of joints and the use and proper application of sealants and other materials are necessary. Where untried details or materials are used, the testing of a mock-up for air infiltration and for water leakage can help in assuring that the proposed details and specified materials adequately exclude air and rainwater from entering the wall. The issues of field installation and field quality control must be addressed by the designer and specification writer. For more complex and innovative systems, specifying quality control specialists for inspecting and monitoring the installation of envelope systems in Section 01450, Quality Control, and specifying that the application only be performed by installers trained, approved, or licensed by the manufacturer will assist in reducing moisture problems in service. Also important are operation and maintenance, both for the envelope and for the mechanical equipment. Face-sealed joints in particular need to be inspected and repaired at regular intervals. If pressurization or depressurization are used to reduce air infiltration/exfiltration, documentation of proper fan settings is critical. However, these concerns are outside the scope of MNL 40 and will not be further discussed.

Moisture Analysis is still an evolving art and science. While great advances have been made in the development of reliable and easy to use models and methods, the availability of some needed input data, such as product specific materials data, is still limited. Nevertheless, even with the existing models and input data, the results of the analysis will provide more, and more reliable, information on the suitability of a particular building envelope design than the sole application of simple prescriptive design rules.

Outline of Manual 40

The following is a brief outline of each of the chapters of MNL 40, Moisture Analysis in Buildings:

Introduction: Outlines the purposes of the manual. Much of the discussion above was extracted, summarized, and adapted from that introduction.

Glossary: By Mark A. Albers[2] A compendium of terms used in moisture analysis, primarily terms used throughout the manual.

Chapter 1: Moisture Primer – by Heinz Trechsel[1]

The Primer is intended to provide an overview of the technology related to moisture movement and condensation in the building envelope for the practitioner and others with only limited or no prior background in building physics. It is not intended to replace more detailed handbooks, or to supply a complete knowledge of building physics.

Terminology: Defines and explains the more important terms used in moisture analysis.

Properties of Air: Describes the relationships of the properties defined in the Terminology section and includes examples of how to determine properties by means of psychrometric charts. Also refers the reader to a CD ROM included in the manual which contains two programs for converting properties of air.

Moisture Sources: Moisture sources are covered in greater detail in MNL 18, this chapter provides some guidance of the type of sources as well as the quantitative contribution of each to the total moisture balance in the building. This section provides a summary of data introduced in much greater detail in Dr. Jeffrey Christian's chapter 8 of MNL 18.

Building Materials: Chapter 3 on Materials provides in detail specific material characteristics, this section provides a brief overview of generic materials and their moisture related properties.

Acceptable Moisture Levels: Discusses the three components of acceptable moisture levels: Health and human comfort of the building occupants; optimization of the conservation of building contents, such as antiquities and artworks in museums; and the maintenance and longevity of the building structure.

Moisture Movement: The chapter discusses mass transport and diffusion and driving forces: Wind pressure, stack effect, mechanical ventilation, and water vapor pressure.

Chapter 2: Weather Data – by Anton TenWolde[3] and Donald Colliver[4]

Chapter 7 on Climate of MNL 18 provided a significant base of weather data from sources such as ASHRAE, the Department of Defense, and from the National Oceanic and Atmospheric Administration. These data are not duplicated in MNL 40. Although that type of data can be useful in moisture analysis, they were all developed for purposes other than moisture analysis, such as for building heating and cooling load calculations and for sizing heating and cooling equipment. The weather conditions described in those data occur rarely, and data of such extremity would seldom be called for in moisture analysis. The chapter discusses the needs for data designed specifically for moisture analysis and develops various suggested approaches.

---

[2] Johns Manville, Littleton, CO.

[3] USDA Forest Services, Forest Products Laboratory, Madison, WI.

[4] Biosystems and Agricultural Engineering Department, University of Kentucky, Lexington, KY.

The chapter provides tables of weather data specifically designed for moisture analysis for 240 United States cities and 85 Canadian cities. The tables give average dry-bulb and dew-point temperatures for the months of January, April, July, and October for use in simplified approximate moisture calculations. The tables are based on measured data taken during the period from 1973 through 1993.

Chapter 3. Hygrothermal Properties of Building Materials – by M.K. Kumaran[5]

MNL 18 contained an entire chapter on moisture related properties as available in the early 1990s. These data only included single point or average values for specific materials. As such, these data are still valid and are not repeated in MNL 40.

Based on new insights and laboratory work, Dr Kumaran's chapter updates and expands this information. The chapter includes recent information from the work of the International Energy Agency Annex 24. Thus, instead of a single value for Water Vapor Permeability, the new tables provide values for ranges of relative humidities and temperatures. Similarly, instead of a single value for Thermal Conductivity, the tables include values for different temperature ranges and, when appropriate, for various levels of density. Reliable values for the Equilibrium Moisture Content for the full range of relative humidity are listed for all common building materials. Recent information on Moisture Diffusivity as a function of moisture content and air permeability can also be found in the chapter. The chapter also lists definitions for all hygrothermal properties and discusses the principles involved in their measurements.

Chapter 4. Failure Criteria – by Hannu Viitanen[6] and Mikael Salonvaara[6]

Although moisture analysis can determine the moisture content of envelope materials, the surface conditions (temperature and relative humidity), and the duration of any excursions, their effect on the materials needs to be evaluated separately. The chapter by Viitanen and Salonvaara will be helpful in such an evaluation. The chapter discusses:

Basic Concepts of Biodegradation

- Environmental factors for development of failures (humidity, temperature, exposure time, materials),
- Failure organisms (mold, fungi, insects, termites, beetles),
- Moisture and microbial problems in buildings,
- The role of materials and the effect of different failure mechanism on wood and gypsum materials.

Critical humidity and moisture level connected with temperature and exposure time are often very important factors for the development of mold and decay. Mathematical models to predict mold and decay are shown and discussed. The critical humidity for mold development is around RH 80 - 95 % and for decay around RH 95 - 100 %, depending on temperature, exposure time, and material. Manual methods and advanced

[5] Institute for Research in Construction, National Research Council of Canada, Quebec, Ont., Canada.

[6] VTT Building Technology, Finland.

numerical methods to predict mold growth are discussed; these models can be attached to the building hygrothermal analyses models discussed elsewhere in this Manual. Estimation, uncertainty and errors of failure predictions, and future prospects are briefly considered.

The chapter also discusses corrosion and corrosive categories and calculation methods to predict failures. In addition, the chapter presents a series of definitions of failure and failure mechanisms.

Chapter 5. Overview of Hygrothermal (HAM) Analysis Methods – by John Straube[7] and Eric Burnett[8]

The objective of this chapter is to provide some background, a brief overview of the various building hygrothermal analysis methods (HAM), and to list and compare analytical procedures, and models.

The chapter also discusses the needs of different groups of people for moisture analysis:

- The building and building envelope designer,
- Those who conduct assessments, such as forensic studies of building failures, and
- Those interested in the performance of specific products, components, or combinations of materials, such as building material producers and suppliers, specification writers, designers and educators.

Chapter 6: Advanced Numerical Models - by Achilles N. Karagiozis[9]

The three manual methods and the two models, MOIST and WUFI described in the following chapters, are useful to the building designer and practitioner. They also serve the building research community. There are, however, advanced models available that are primarily directed toward the research community. This chapter discusses the concept of moisture engineering, introduces a classification of Hygrothermal models as developed by the International Energy Annex 24, further extends this classification, and describes the theoretical background of advanced models.

One distinction of advanced models is that they include not only condensation transport mechanisms, but also air leakage, rainwater penetration, and phase changes. They thus model the actual conditions within the building envelope with greater precision than the more simple models. However, they also have to rely on accurate data regarding air leakage, air leakage sites, and rainwater penetration seldom available to the designer. In addition, some of these advanced models also require mainframe computers. Advanced models are not normally used by building practitioners, they are, however, indispensable tools for researchers to develop general theories and guidelines for designers, building material and component manufacturers, and for those charged with preparing codes and standards for building constructions.

---

[7] Building Science Consultant, Waterloo, Ont., Canada.

[8] Departments of Civil and Environmental Engineering and Architectural Engineering, Pennsylvania State University, University Park, PA.

[9] Building Thermal Envelope Systems and Materials, Oak Ridge National Laboratory, Oak Ridge, TN.

The chapter provides an elaborate account of all the requirements needed to qualify as an advanced hygrothermal model. The integration of the envelope models with indoor air quality models to accurately describe the holistic performance of buildings is also presented.

The chapter summarizes 11 advanced models:

- WUFI, by Hartwig Kuenzel;
- MOISTURE-EXPERT, by Achilles Karagiozis;
- SIMPLE FULUV, by Øyvind Økland;
- TRATMO2 (Mikael H. Salonvaara;
- TCCC2D, by Tuomo Ojanen;
- HMTRA, by D Gavin and B.A. Schrefler;
- DIM3.1, by John Grunewald;
- 2DHAV, by Arnold Janssens,
- LATENITE, by Achilles Karagiozis and Mikael Salonvaara;
- FRET, by M. Matsumoto, S Hokoi, and M. Hatano; and
- FSEC 3.0, by Muthusamy V. Swami, Lixing Gu, and Philip W. Fairey.

For each model, the chapter provides a one to two page summary providing its theoretical basis, explaining the distinguishing features, and showing examples of graphic outputs.

Chapter 7: Manual Analysis Tools – by Anton TenWolde[3]

This is an updated version of Chapter 11, Design Tools, in MNL 18. The chapter describes three steady-state methods that are useful in determining the potential for condensation on the surface of building envelope layers under stated conditions. If condensation is likely to occur under extreme, but realistic conditions, further analysis using computer models is indicated.

Because the manual methods are based on selected conditions, they do not provide any measure of the duration of potential condensation. Since the potential for deterioration and failure is dependent on the duration of moisture excursions, steady-state methods are not suitable by themselves for risk analysis. Also, the three methods only consider moisture transfer by diffusion and do not include transfer by air movement or rainwater leakage.

The three methods discussed are the Dew Point Method, the Glazer Diagram, and the Kieper Diagram. All three methods compare vapor pressures within the envelope, as calculated by vapor diffusion equations with saturation pressures, based on temperatures inside the envelope. If the calculated vapor pressure is above the saturation pressure at any point within the envelope, condensation is indicated.

The three methods are described and examples and solutions are provided. Included are Tables for Saturation Water Vapor Pressures as required for conducting the analysis.

Chapter 8: Numerical Method For Design: Model MOIST - by Douglas M. Burch[10] and George Tsongas[11]

MOIST was one of the earliest models developed for the building practitioner and the researcher to predict the one-dimensional transfer of heat and moisture in walls, cathedral ceilings, and low-slope roof constructions. It predicts the temperature and moisture content of each construction layer and the relative humidity at the construction layer interface, as well as the moisture and heat transfer fluxes at both the interior and exterior envelope faces, as a function of the time of the year. The model can also be used to predict the annual variation of indoor relative humidity.

The model thus allows the designer to determine whether a particular envelope design might lead to elevated moisture and for what duration, thus allowing an estimate of the risk of decay and failure, or whether the installation of a vapor retarder is indicated. MOIST also can predict the surface relative humidity, thus allowing an estimate of the potential for growth.

The chapter outlines the model theory, discusses the limitations: It is one-dimensional and thus does not include the effect of framing members, two and three dimensional effects, such as vertical movement of moisture. The model does also not include exterior wetting of the wall by rain or rainwater penetration and the moisture transfer by air movement. However, the model has proven useful for providing the type of information most designers require, is easy to use, and does not require inputs that are either difficult or impossible to get at the design phase of a building. The chapter describes the model, provides illustrative examples, shows a preview of the dialog boxes that the user will see and sample printouts of results.

The program has been verified extensively at the National Institute of Standards and Technology.

The MOIST program itself is provided on a CD ROM included at the end of the new manual. The disk also includes the user's manual. Thus, the reader can, with some practice, conduct moisture analysis as needed for his or her building envelope designs.

Chapter 9: WUFI ORNL/IBP – A Hygrothermal Design Tool for Architects and Engineers – by H. M. Kuenzel[12], A. N. Karagiozis[10], and A. H. Holm[12]

The second model described in detail is a design oriented tool evolved from an earlier model of WUFI, prepared by Hartwig Kuenzel. The WUFI ORNL/IBP version of the model was developed to be more user friendly and to include inch/pound units for ease of use by American architects and Engineers.

The chapter describes in detail the concept of the model. In separate sections, the chapter covers Moisture Storage, Moisture Transport (vapor diffusion and liquid water transport), Material Properties, Boundary Conditions, Indoor and Outdoor Climate, and the limitations of the program. WUFI has been tested rigorously. However, as with any model, the results are only as reliable as the input data. The chapter provides some common sense suggestions for minimizing such errors.

---

[10] Heat and Moisture Analysis, Inc., Olney, MD.
[11] Ph.D., P.E., Consulting Engineer, Portland, OR.
[12] Frauenhofer Institute für Bauphysiks, Holzkirchen, Germany.

The chapter also lists some limitations of the model. Similar to MOIST, WUFI ORNL/IBP only deals with one dimensional processes. It does not include moisture transport by air movement and by rainwater leakage. It does, however, include wetting of the exterior skin by rainwater. (Moisture transport by air and rainwater can be treated by the two dimensional model WUFI 2d which is not described in this manual.)

The chapter provides examples of problems which can be solved with WUFI ORNL/IBP and contains figures showing output data.

The WUFI ORNL/IBP program itself is provided on a CD ROM included at the end of the new manual. The disk also includes the user's manual. Thus, the reader can, with some practice, conduct moisture analysis as needed for his or her building envelope designs.

Chapter 10. A Look to the Future - by Carsten Rode[13]

The technology of building moisture analysis is still evolving. It was therefore deemed appropriate to include a chapter that identifies tendencies that will affect the future practice of moisture analysis:

Weather data need to be refined and restructured to be more useful for moisture analysis. To include rainwater penetration in future models, additional data on wind driven rain must be developed. There also needs to be developed a moisture design reference year specifically for moisture calculations instead of for energy calculations (Chapter 2 is a first attempt at developing such a reference year). Contemporary media and data bases are expected to be used to gather and present existing and new data so they become more widely accessible. Weather data will also be expanded considerably; the necessary research and testing is already underway in many locations and countries.

Together with material data, additional information is needed on failure criteria. Similarly, additional data is needed on durability of materials in a life cycle perspective.

Current analysis methods suitable for building designers and practitioners are one-dimensional models. To overcome some of the shortcomings, multidimensional models will be developed.

A major effort will be made in the area of benchmark testing, dissemination, and integration of models into the design process.

Whole building modeling will become a practical and economical way to optimize not only moisture performance, but the integrated performance of the entire building.

Equipment and instruments for in-situ moisture measurements and moisture alert systems to be integrated into the building structure will provide for monitoring of the moisture performance during service and will allow timely intervention where threshold moisture content and relative humidities are exceeded.

Finally, new construction methods and materials will be developed that will provide for more moisture-resistant constructions.

---

[13] Department of Buildings and Energy, Technical University of Denmark, Lyngby, Denmark

References

[1] Building Environment and Thermal Envelope Council is a Council of the National Institute of Building Sciences, Washington, DC.

[2] Bales, E. L., et al. "BETEC Moisture Analysis Tutorial," *Water Problems in Building Walls: Evaluation, Prevention, and Repair, ASTM STP 1352*, J. M. Boyd and M. J. Scheffler, Eds., American Society for Testing and Materials, West Conshohocken, PA, 1999, pp. 91-114.

[3] *Moisture Control in Buildings*, Heinz R. Trechsel, Editor, ASTM Manual Series MNL 18, American Society for Testing and Materials, Philadelphia, 1994.

[4] Conference on Condensation Control in Dwelling Construction, Housing and Home Finance Agency, May 17&18, 1948.

[5] HUD Minimum Property Standards for One- and Two-Family Dwellings, 4900.1,1980 (latest edition).

[6] ASHRAE Handbook of Fundamentals, American Society of heating, Refrigerating, and Air-Conditioning Engineers, Atlanta, 1993.

[7] Lstiburek, J. and Cormody, J., "Moisture Control for New Residential Buildings," *Moisture Control in Buildings, MNL 18*, H. R. Trechsel, Ed., ASTM International, West Conshohocken, PA, 1994.

[8] Anderson, L. O. and Sherwood, G. E. "Condensation Problems in Your House: Prevention and Solutions," *Agriculture Information Bulletin No. 373*, U.S. Department of Agriculture, Forest Service, Madison, 1974.

[9] *ASHRAE Handbook of Fundamentals*, American Society of Heating, Refrigerating, and Air-Conditioning Engineers, Atlanta, 1997.

[10] Odom, J. D. and DuBose, G. *"Preventing Indoor Air Quality Problems in Hot, Humid Climates: Design and Construction Guidelines,"* CH2M HILL and Disney Development Corporation, Orlando, 1996.

[11] Lstiburek, Joseph, "*Builders Guide to Hot-Humid Climates,*" Building Science Corporation, Westford, 2000.

SECTION V

Claire Benge[1]

A Verification Method for Prevention of Penetration of Moisture to Prove Compliance of Performance-Based Building Codes

---

Reference: Benge, C., **"A Verification Method for Prevention of Penetration of Moisture to Prove Compliance of Performance-Based Building Codes,"** *Performance of Exterior Building Walls, ASTM STP 1422,* P. G. Johnson, Ed., ASTM International, West Conshohocken, PA, 2003.

Abstract: The New Zealand building code is performance-based, and to facilitate proof of compliance the Building Industry Authority publishes Approved Documents that provide tests or calculation methods (Verification Methods) and "cookbook" solutions (Acceptable Solutions).

Currently the Acceptable Solution for the prevention of penetration of moisture into a building that will cause undue dampness or damage to the building deals only with generic materials such as timber weather boards, brick veneer and stucco on light timber framing.

The Verification Method contains test methods to be carried out in a laboratory. Not only can such tests be costly, but they may also not take account of site conditions or of the protection provided by the layout and orientation of a building.

This paper will examine the possibility of an additional Verification Method that will show compliance with the performance based building code by dealing with the adequacy of the building envelope in terms of such issues as its materials and systems, the joints within and with other claddings, any secondary defense methods, disposal of moisture that does penetrate, and exposure and durability considerations.

It is hoped that the result of this and further studies will be a Verification Method for the building envelope useful for manufacturers to predict that their product will comply with the code and for designers who are dealing with unique situations to prove compliance without costly testing.

Keywords: performance based building codes, building envelope, claddings, moisture

---

[1] Architect and Technical Adviser, Building Industry Authority, P O Box 11846, Wellington, New Zealand.

Introduction

The building industry in New Zealand is suffering from a crisis similar to that in Canada and the United States of America commonly known as the "leaky condominium" crisis. In New Zealand it is not confined to condominiums, as we do not build them to the same scale as in North America, but the same problem of leaking buildings is being experienced throughout the range of light timber construction frequently clad in plaster based systems, commonly used in New Zealand for dwellings, both single and multi-unit, and for smaller commercial construction.

The Claddings Institute of New Zealand (CINZ), with a membership consisting of a cross section of the building industry including manufacturers, designers, building officials, building surveyors[2] and academics, held a forum to discuss the issues. Consensus at this forum laid the blame on no single reason for the crisis but a collection of reasons, including poor detailing at the design stage, poor training, workmanship, and control, with no one person or organization on site willing to take responsibility. Although the forum concluded that the existing building controls do cover the requirements for weatherproofing of buildings, more work could be done to provide easy-to-understand Approved Documents so that building officials, designers and builders have quick and more flexible time saving solutions readily available.

Prior to this Forum, the Building Industry Authority (BIA) had already recognized that need and had planned or commenced several projects that would contribute to achieving that purpose. These included assisting the Building Research Association of New Zealand (BRANZ) to develop standard detailing for junctions in claddings, writing a new Standard for building papers to cover specific New Zealand construction methods and changes in materials, and extending the Acceptable Solution for clause E2 "External Moisture" (E2) of the New Zealand building code (the building code) [*1*] to cover the types of claddings that had become popular in recent years.

It was the development of this extended Acceptable Solution that generated the idea to have a verification method that is not a test method but rather a method of proving that cladding materials comply with the building code by confirmation that all conditions have been considered and dealt with. In order to understand what is meant by a verification method, it is essential that the New Zealand building regulatory system be explained for those to whom it is new.

An Overview of the New Zealand Regulatory System

The legislation

The New Zealand Building Act 1991 [*2*] introduced the Building Regulations 1992 [*3*] and a 6 months lead-in period saw the full introduction of the building code as the first schedule of the Building Regulations in January 1993. This national building code, which is performance-based, was introduced to give consistency to an industry that was

---

[2] A Building Surveyor in New Zealand is a professional person who provides reports on the condition, suitability of buildings and remedial work necessary. The position does not require a degree in NZ; they are usually engineers, builders or drafters.

previously governed by a myriad of acts, regulations and local by-laws riddling the country's building regulatory regime.

Performance-based building codes

Traditionally, building regulations have been prescriptive, saying what must be done and how to do it. By their nature such prescriptive systems of governance are very restrictive because it is unlikely they can cover every foreseeable circumstance and certainly they cannot cover unforeseeable circumstance. In recent times, as regulatory authorities search for more flexible, less time consuming and cheaper methods of regulating building, they have been leaning towards performance-based building codes. Performance-based building codes state what must be achieved, not how to achieve it, and are therefore by nature very broad and versatile.

The New Zealand building code

The building code has 37 clauses covering all aspects of building. These are the mandatory clauses and are performance-based. Each clause has a 3-tiered system of sub-clauses as follows:

1. Objective – which states the purpose of that clause, usually based on health, safety and accessibility
2. Functional Requirement – which states what has to be achieved,
3. Performance – which states how to achieve the Objective in either a qualitative or quantitative form.

Proof of compliance with performance-based building codes

Compliance with performance-based building codes can be made easier to prove for commonly used construction by provision of an approved prescriptive method of compliance. In New Zealand prescriptive based documents (the Approved Documents) have been produced as a means of compliance with the building code. The Approved Documents consist of Verification Methods, which are usually calculations as are done for structural design or for tests, and Acceptable Solutions, which are "cookbook" methods.

The Approved Documents are not mandatory, as can be seen from the model shown in Figure 1, and owners or designers are entitled to use alternative solutions that must show compliance with the performance requirements of the building code. The most common proof that an alternative solution complies with a particular clause is to demonstrate that the material, system or design method is similar to the Approved Document for that clause. If the proposed alternative solution is not sufficiently close to an Approved Document, proof that the proposed alternative solution will comply with the Performance criteria must be produced. The Territorial Authority, through its building officials, makes that decision, being required under section 34(3) of the Building Act, to grant the consent if it is satisfied on reasonable grounds that the provisions of the building code would be met. Requirements for "proof" and "reasonable grounds" can only be determined on a case-by-case basis.

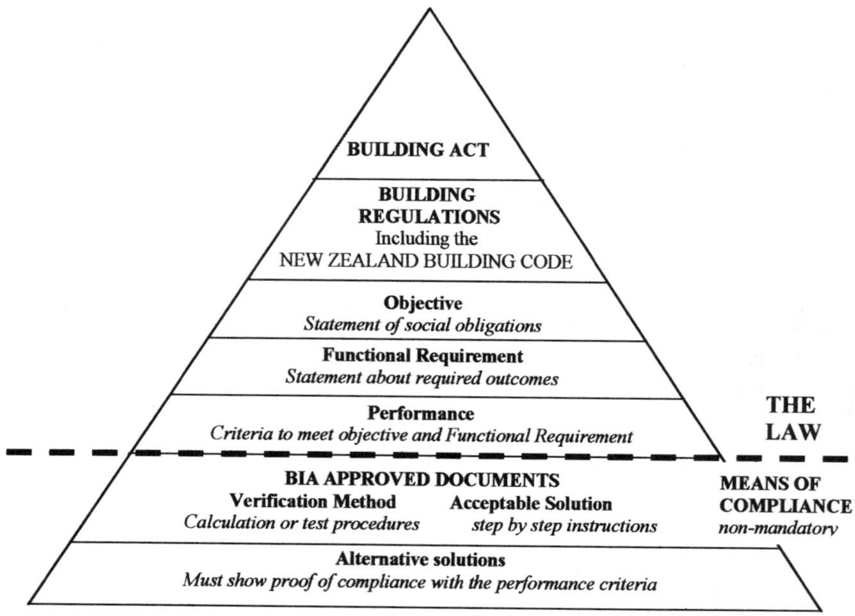

Fig. 1 Model of the New Zealand Building Regulatory System

The decision to approve alternative solutions is not easy, so that territorial authorities often restrict approval of alternative solutions to senior building officials who have the experience and training required. There is no formal qualification for building officials, although one is under development. Training for assessment of alternative solutions, and for verifications such as is proposed in this paper will need to be included in such a qualification.

Clause E2 "External moisture"

The clause of the building code relevant to the prevention of penetration of moisture is clause E2. The Objective of E2 is *to safeguard people from illness or injury which could result from external moisture entering the building*[*1*]. Under the Functional Requirement buildings are required to be constructed *to provide adequate resistance to penetration by, and the accumulation of, moisture from the outside*[*1*]. The Performance criteria to meet the Objective and Functional Requirement is that *roofs and exterior walls shall prevent the penetration of water that could cause undue dampness, or damage to building elements*[*1*].

It is only prevention of *undue* dampness or damage that is required by E2. In other words, a little dampness or damage is allowed. It becomes "undue" when moisture penetrates or accumulates to cause dampness that would support mould or fungal growth, or cause damage to building elements behind the building envelope. It is "undue" if clause E3 "Internal moisture" (E3) or B2 "Durability" (B2) are not complied with. Ingress of moisture is therefore acceptable if it is infrequent and able to dry out quickly.

It is therefore worth noting that the building code does not expect a building to be totally waterproof all the time, but rather that the unusual storm (outside the 50 year rate of return predicted) of higher winds, heavier rain or of longer duration does not need to be catered for.

Verification Methods

As mentioned above, Verification Methods usually consist of a calculation method, such as SNZ 4203:1992 "Code of practice for general structural design and design loads for buildings", [*4*] used to design the structure of a building, or by tests such as AS/NZS 4284:1995 "Testing of building facades" Appendix D Water Penetration Testing [*5*], used to test claddings in laboratory conditions or on site.

Two of the 37 clauses of the building code have a different type of Verification Method that is perhaps best described as a "complex checklist". This method relies partly on the abilities of the user of the checklist, since some skill and knowledge of the subject is required.

Clause B2 is one clause that has a checklist for a Verification Method (B2/VM1) [*6*]. B2 states that all building elements shall meet a required durability. There are four main considerations used to define consumer expectations to determine whether the durability requirement should be the life of the building (being not less than 50 years), 15 or 5 years. These are:

- Ease of access
- Ease of replacement
- Detection of failure
- Structural role

Predicting the durability of materials is a relatively new science. Because it would be impossible to conceive all the properties of every material, the checklist cannot be inclusive or it is of limited use. In B2/VM1 proof of performance can be shown by one or more of the following:

1. In-Service History
2. Laboratory Testing, or
3. Comparison with Similar Materials

Each of those has a more detailed checklist of its own. For example verification of durability by In-Service History must take into account the length of service, the environment and intensity of use, any reaction with adjacent materials, limitations in performance, degree of degradation, and changes in formulation within that in-service history, or any other matter that is likely to affect the durability of the building element under consideration.

A highly developed knowledge of how materials react and what is likely to cause deterioration is required by the user of such a verification method, so that the Territorial Authority when checking the proof of durability would need to rely on the reputation and skills of the user, either an experienced materials scientist working for a certified laboratory, or a designer who coordinates available information. Product appraisal certificates such as produced by BRANZ are commonly used here.

Clause F1 "Hazardous agents on site" (F1) is the other clause that has a checklist for a Verification Method (F1/VM1) [*6*]. F1 requires that sites be assessed to determine the presence and potential threat of any hazardous agents or contaminants. Like the durability requirements for B2, it is not always possible to predict the presence of likely contaminants on site or what they will be. F1/VM1 requires that a site be evaluated by studying the site history, visually surveying the site and where indicated, undertaking further investigation to identify any hazardous agents or contaminants and evaluating the risk in relation to the proposed building.

Why have a new Verification Method for clause E2?

There is an existing Verification Method for clause E2 (E2/VM1) [*6*]. This calls up two different tests as a means of showing compliance with clause E2 3.2 of the building code. Both test methods are generally carried out in a laboratory, and usually only on a small part of the whole building envelope, either a unit such as a window or skylight, or a small area of wall or roof cladding. The tests are designed for testing the manufacturer's product, be it a window, skylight, roof tile or wall cladding but can be used when developing designs for large buildings. They can be but are not usually used to test the junctions between different parts of the building envelope such as window to wall cladding, skylight to roofing. This is unfortunate because junctions are the places in which it is most likely that leaks will occur. The existing test methods therefore are not giving the industry the information it most needs. Additionally, laboratory testing can not always take site conditions into account.

Manufacturers of windows usually supply some construction details in their technical information. Manufacturers of claddings also supply details of their claddings with the various window and door profiles, and even for small penetrations such as pipework. Neither groups of manufacturers can afford to test all the variations and permutations likely to be used with the large variety of cladding materials, and door and window profiles available in New Zealand.

Additionally, as architects and designers try to be innovative with the available materials in order to follow fashion, cultural or philosophical trends in the design of buildings, new details are designed, usually to be tested on site, post-completion, with disastrous consequences if even a minor detail is incorrect.

Thus there are at least three strong reasons for having such a verification method:

- Reducing the complexity of proving compliance,
- Reduce the cost of proving compliance, and
- Encouraging innovation.

Initial development

A workgroup of people working within the building industry was convened by the BIA to advise on extending the existing Acceptable Solution to include claddings so far not dealt with, but which are becoming so commonly used that they should be included. (Members of the workgroup are acknowledged at the end of this paper.) The current Acceptable Solution includes only the most commonly used generic claddings, namely timber weatherboards, brick veneer and stucco on timber framing. These three cladding materials have been traditionally used over New Zealand's short history but new proprietary claddings have become so popular in recent years that fibre cement is being used rather than timber for weatherboards and high build plasters over Exterior Insulation and Finish Systems (EIFS) and fibre cement are being used instead of the traditional stucco.

In the process of discussing a brief for the writing of an Acceptable Solution, the workgroup developed a list of methods by which a cladding deals with moisture penetration or collection of moisture within the cladding. The following list of methods (M) is a modification of that produced by the work group:

M1. Inflow of moisture by:

(a) Deflecting water before it reaches the cladding, such as:

- Eaves or verandahs (width and height)
- Balconies (width and height)
- Dripline (not always appropriate in very high wind areas)
- Flashings that project eg head flashings,
- Rough surface to break up and bounce off moisture eg concrete or plaster

(b) Shedding water from the surface by having an impervious material or a repelling surface, such as

- Metal or plastic claddings,
- Protective coatings and paint systems,
- Silicone repellent coatings

M2. Outflow of moisture – the way it leaves the cladding system:

(a) Drainage by:

- Cavity – such as in brick veneer, non-rigid-backed stucco
- Weepholes – such as in the bottom of brick veneer
- Underlay – as a secondary means of defence behind the cladding,
- Drainage medium
- Drained joints
- At interfaces – such as between stucco and cellulose cement sheet
- Weather or anti-capillary grooves

(b) Deliberate drying (evaporation) by:

- vapour permeable external surface
- Internal ventilated cavity
- Absorbent backing material

The workgroup also contended that the overall performance of the cladding must be considered within the context of the building, its location and additional building code requirements of the cladding. A list of the relevant contexts (C) includes the following:

C1. Exposure to:

- Wind,
- Rain,
- Sun,
- Corrosion (sea spray, geothermal or industrial),
- Site specific wear and tear from likely loads.

C2. Protection

- No protection

- Protected by paint finish or protective coating
- Protected by position (eaves, verandah, deck etc)

C3. Additional building code requirements

a) Durability requirements
  - Ease of access and replacement
  - Detection of failure
  - Normal maintenance
  - Structural role

b) Fire requirements

c) Sound requirements

While forming an excellent basis for a brief to write an Acceptable Solution in which the "cook book" solution had to cover the above points, the workgroup concluded that it could also be an excellent base for a verification method, that would confirm that all conditions have been considered and dealt with; in other words, the "complex checklist" method described in the introduction.

It is unlikely that any one method on its own could be considered to be sufficient to prevent undue dampness or damage. In a paper entitled "Designing for Durable Wood Construction: The 4 Ds", Hazleden outlines a simple set of basic principles used to cover such a complex set of solutions. These principles consist of "Deflection", "Drainage", "Drying" and "Decay resistance". While advocating that all 4 of the "Ds" are necessary to ensure a completely weathertight building, Hazleden also points out that "Deflection" can cope with more than 90% of incident rain given a properly designed deflection system. A properly detailed "Drainage" system can theoretically deal with the remaining 10% of water, reducing the amount of water entry to less than 1%, which could be considered reasonable.

The fourth "D", "Decay resistance", is not so much a solution for prevention of penetration of moisture but more of a damage control after the act, potentially preventing "undue" damage in timbers but not dealing with any "undue" dampness on other materials such as linings that may not deteriorate but could develop unhealthy mould. Thus relying on the use of decay resistance as a back up to the other "Ds" might not ensure full compliance with the building code.

Further Development of the Verification Method

The list developed by the workgroup is really only the beginning. To have a verification method that could predict whether moisture could cause undue damage to a building, each method by which a building deals with moisture penetration or accumulation within the cladding in the first section has to be considered within the context of the building as listed in the C1, C2 and C3 above.

Hazleden's observations that deflection deals with at least 90% of rain and drainage deals with nearly 10% indicate that different methods must be given different weightings in terms of the effectiveness of the complete design. A design could concentrate on deflection (wide verandah all around the building) and not need to deal with the other 3 "Ds". However, in such a design, exposure to winds would need to be a factor in a country such as New Zealand where strong winds can sometimes cause falling rain to develop a near horizontal incidence. On the other hand, because 90% of the normal defence is carried out by deflection, a design that relied on drainage and drying would need to be very rigorous in order to ensure weathertightness.

A proposed Verification Method would therefore be a complex matrix, taking into account the means of dealing with moisture penetration or accumulation by a combination of preventing inflow of moisture and directing it out while taking into consideration exposure, protection and additional code requirements. The detailed matrix may be too complex to put into a table format but a summary or check-sheet for each product could be an appropriate form to include that would show the Territorial Authority what has been considered.

Knowledge gaps

Work done by the BIA to extend the existing Acceptable Solution for E2 (E2/AS1) [*6*] and by BRANZ to develop a set of details of "Junctions of the Building Envelope" [*8*] has shown that there are some major gaps in knowledge of how claddings prevent the penetration of moisture or its accumulation in the building envelope. Detailing of claddings has tended to develop by the trial and error method, something of a hit-and-miss system at the best of times because it relies firstly on progressive amendments over a period of time, secondly on designers admitting when they have made mistakes, and thirdly on dissemination to the industry of the information gained from trial and error.

Building papers as underlays and wraps

The traditional role of building paper behind claddings as well as the basic materials from which it is made has changed in New Zealand. New papers are being used, either because of their fire-retardant qualities or because of their strength. Strong, translucent polyethylene papers are primarily being used to wrap around the timber framing, providing a dry, wind-free, light interior for finishing construction of the framing, and only secondly as an underlay for the cladding. Research must be done to define the role of traditional building paper before confirming that it is appropriate to use such paper as a wind barrier. Full understanding of the new building wraps is needed before we can be sure that they can function in the traditional role of building paper as well as their role in the construction process.

Building papers as barriers to wind pressure

The use of light weatherboards such as aluminium, uPVC[3] and cellulose cement has seen the increased need for underlays to act as a wind barrier in the higher wind areas as well as performing their traditional moisture barrier role. Their lightness usually means flexibility which means more water and wind penetrates the primary screening they provide. Theory based on rain screens and pressure equalisation shows that such a wind barrier prevents wind pressure from pushing moisture through the building paper underlay. This however moves the pressure to the junctions between the cladding and

---

[3] Unplasticized polyvinylchloride

window/door penetration. BRANZ has temporarily solved the problem by suggesting air seals in the gap around windows and doors in high and very high wind areas. These seals are more of a trial and error solution than one based on physical tests, but it is now possible to do some research to show how and why it works, and to enable some fine tuning to be applied to such solutions. These types of solutions must additionally be shown to be vital or a cost conscious industry will ignore and omit them

Miscellaneous

Paint finishes and repellent coatings, solar driven moisture, geo-textile drainage media, geothermal and industrial corrosion, durability and efficient placing of sealants are all aspects of moisture in which there are large gaps in our knowledge. Some will remain experimental used as an alternative solution rather than being incorporated in an Approved Document. Others can be adopted after consideration of existing knowledge and careful development as answers emerge. A verification method will help in dealing with these problems by demanding answers, particularly during product development, as they are needed to create a tight and reliable solution to the building industry's moisture entry problems.

Conclusion

A verification method that shows compliance with the performance-based building code by dealing with the adequacy of the building envelope through the use of a "complex checklist" has potential. It could be of large benefit to the building industry because laboratory testing is costly, is limited in the variation of detailing it can cover, and cannot always take into account site conditions.

As the issue of moisture entry is a fairly complex one, because the physics of water penetration and accumulation in buildings is not an exact science, research will be needed to improve confidence in such a verification method.

Acceptable Solutions for the building code are by their very nature conservative. Verification methods do not need to be conservative but can be used to produce innovative design. If the necessary research is adequately carried out, a verification method such as is discussed could allow less conservative, but fully compliant, designs to be used, freeing up the industry resources of time and money used to prove alternative solutions, the more costly route to compliance with the building code.

With BRANZ already working on some of the suggested research, and the industry involved in discussion about solutions to the leaky building problem, it is possible that such a verification method could be developed within two years in an initial, necessarily conservative, form. As knowledge and experience develop, it could become a very flexible and useful tool in ensuring that our buildings do not suffer from "undue" dampness or damage due to the penetration and accumulation of moisture in or through the building envelope.

Acknowledgments

My thanks to the following people who were members of the BIA workgroup: Phillip O'Sullivan, Engineer, Prendos Ltd., Auckland; Rose McLaughlan, Building Certifier, A1 Certifiers Ltd., Drury; Don Bunting, Architect and General Manager, Construction Information Ltd., Auckland; Joe ten Broeke, Senior Technical Adviser, BRANZ; and Katharine Wheeler, Architect and Building Official, Wellington City Council, Wellington.

The ideas expressed in this paper are the author's and do not necessarily represent the views of the workgroup.

References

[1] The Building Code (being the First Schedule of the New Zealand Building Regulations 1992), published under the authority of the New Zealand Government, Wellington, New Zealand.

[2] The New Zealand Building Act 1991, published under the authority of the New Zealand Government, Wellington, New Zealand.

[3] The New Zealand Building Regulations 1992, published under the authority of the New Zealand Government, Wellington, New Zealand.

[4] Standards New Zealand, SNZ 4203:1992, "Code of practice for general structural design and design loads for buildings," Wellington, New Zealand.

[5] Standards Australia and Standards New Zealand, AS/NZS 4284:1995 "Testing of building facades" published jointly by Standards Australia, Homebush NSW Australia and Standards New Zealand, Wellington, New Zealand.

[6] Building Industry Authority (New Zealand), 1998, The Building Code Handbook and Approved Documents published by Standards New Zealand, Wellington, New Zealand.

[7] Hazleden, D. G. and Morris, P. I., "Designing for Durable Wood Construction: The 4 Ds", *CIB 8th International Conference on Durability of Building Materials and Components: 1998.*

[8] Building Research Association of New Zealand, "Details for Junctions and Openings in Wall Claddings," published by BRANZ, Judgeford, Porirua, New Zealand, 2001.

Francesco J. Spagna[1] *and Stephen S. Ruggiero*[1]

Stucco Cladding – Lessons Learned from Problematic Facades

Reference: Spagna, F. J. and Ruggiero, S. S., **"Stucco Cladding – Lessons Learned from Problematic Facades,"** *Performance of Exterior Building Walls, ASTM STP 1422*, P. G. Johnson, Ed., ASTM International, West Conshohocken, PA, 2003.

Abstract: The authors have investigated many problematic stucco facades around the United States and have documented a number of disturbing trends in the design and installation of stucco that have compromised its reliability. These trends have led to severe system failure, including stucco cracking, leakage to the building interior, deterioration of structural components, and mold growth on building materials within the wall cavity.

This paper examines the current design basis for stucco-clad walls, which for the most part is established by accepted standards of practice within the industry, with only minimum and, in many cases, inadequate standards set by the governing building codes.

Based upon our review of codes and industry literature, and the lessons learned from case histories, we present recommendations for improving design and installation details for conventional stucco.

Keywords: asphalt-saturated felt, brown coat, building paper, cladding, control joint, cracking, curing, diamond mesh, embedment, expanded metal lath, expansion joint, finish coat, flashing, keying, kickout, kraft-paper, lath, metal diverter, paper-backed lath, peel-and-stick membrane, plaster, portland cement, rake flashing, rubberized-asphalt, scratch coat, self-furring lath, sheathing, stucco, veneer, waterproofing membrane, weep screed, window

Stucco – Overview

The three major types of plaster are portland cement, lime, and gypsum. Gypsum plaster has historically been used in interior applications, since it deteriorates with exposure to moisture. Prior to the introduction of portland cement in the United States in the late 1800s, lime was the cementitious binder for most exterior and interior plasters. Portland cement led to a harder, more durable plaster, today referred to as stucco.

[1] Staff Engineer and Principal, respectively, Simpson Gumpertz and Heger Inc., 297 Broadway, Arlington, Massachusetts, 02474.

Stucco has been used to clad residential and low-rise buildings for hundreds of years, with apparent success and acceptance by many building owners. The ability of stucco to imitate more elaborate, expensive claddings such as stone made it a very attractive building material. In the early part of the twentieth century in the United States, exterior portland cement plaster applied to metal lath became a common cladding for residential wood-framed structures. In the later half of the century, its use spread to commercial construction.

Stucco remains a popular cladding of choice in the South, Southwest, and California, and is often found on low- to mid-rise apartments and condominiums, institutional facilities, and municipal buildings; it is not uncommon to find stucco-clad buildings that are 5 to 10 stories tall. Unfortunately, too many stucco-clad buildings exhibit signs of distress, and problems appear within the first few years of service. The increasing tendency of the industry to consider stucco a "surface-sealed barrier wall" as opposed to its more appropriate usage as a drainage wall, typically leads to problematic facades.

Drainage vs. Barrier Walls

The most fundamental reason for performance problems with stucco cladding systems is the failure to achieve a functional drainage system behind the stucco veneer. This results from design and construction practices that fail to provide a watertight back-up layer (membrane), flashings and, most importantly, sufficient weep capacity for the wall area.

Although stucco can be formulated to provide a dense veneer that resists water penetration, it cannot be expected to perform as a "surface-sealed barrier wall." Water will penetrate at cracks, control joints and window surrounds in sufficient quantity as to require a back-up line of defense (i.e., the back-up waterproofing system). The stucco serves as the first line of defense by shedding most of the water that strikes the wall surface, but enough water will penetrate to cause significant leakage and damage if the back-up system is not properly designed and constructed. Industry publications[1-4] offer some guidance on drainage design, but material requirements for the back-up membrane are too general and permit the use of relatively thin paper-based products and the design requirements for weeping multi-story buildings are confusing at best.

Some standards require that stucco surfaces be provided with "sufficient slope" to prevent the accumulation of water, snow, or ice. Stucco should not be used on sloped surfaces or for roofing, no matter how small the roof; this includes window shelves and other flat surfaces such as the tops of parapets. Cracking of the stucco is inevitable on exposed, horizontal surfaces, particularly near the re-entrant corners of windows, and water will accumulate on the flat, back-up waterproofing and eventually leak to the interior.

At parapets, stucco should terminate beneath watertight metal parapet caps. The parapet cap should slope to the roof, be free of penetrations (fastened with concealed hook strips), and have watertight transverse joints preferably flashed with EPDM. If stucco is installed on both sides of a parapet, the back-up waterproofing needs to be continuous up and over the parapet and properly integrated with roofing materials.

Typical Stucco Wall Components

Lath

Modern stucco is generally applied to metal lath, which can be galvanized or stainless steel, and is available in different geometries, including expanded metal lath (diamond mesh), woven wire fabric, and welded wire fabric. Most laths are self-furring; the metal is dimpled or crimped intermittently to provide a 1/4-in. furring space. The furring space facilitates mechanical keying of the stucco with the lath, and serves to protect the back-up waterproofing from the abrasive surface of the lath. Care must be taken to ensure that self-furring dimples are not flattened during lath installation.

The lath is installed over a waterproofing layer, typically a kraft paper and asphalt laminate or an asphalt-saturated felt, which are often referred to collectively as building paper. Lath with a factory-adhered paper backing is also available, in which case the lath and building paper are installed simultaneously. Stucco is typically installed over exterior gypsum or wood sheathing, which is fastened to steel or wood stud walls (Figure 1). Building codes also allow stucco on metal or paper-backed wire lath to be installed directly over steel or wood stud framing, without the use of sheathing, in which case the lath, paper and stucco span the space between studs.

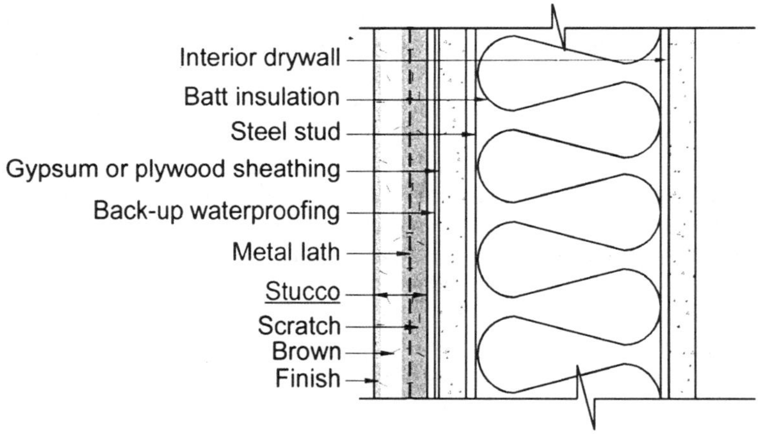

Figure 1 – *Typical Stucco Wall System*

Installation of paper-backed lath over open stud framing often results in unembedded lath, even in the case of self-furring lath (Figure 2). Codes and standards require installing scratch coat with "sufficient pressure" to achieve mechanical keying with the lath. Without sheathing to provide a reactive force, lath deflects into stud cavities under pressure. We have found that keying is often not achieved, resulting in large areas of unembedded lath. The unembedded lath is directly exposed to water flowing on the building paper, which in many cases has led to premature corrosion of the lath.

Figure 2 – *Unembedded, Self-furring Lath Over Open Stud Framing*

Building Paper

The building paper serves as the waterproofing membrane when the wall is in service, but also keeps the substrate dry during stucco application. The Federal Specification for Building Paper, UU-B-790, grades building paper according to water resistance as follows:

- Grade A (high water-vapor resistance, 24-hour water resistance)
- Grade B (moderate water-vapor resistance, 16-hour water resistance)
- Grade C (8-hour water resistance)
- Grade D (water-vapor permeable, 10-minute water resistance)

It is important to note that this specification applies to building papers that consist of pulp fibers. As we will discuss in our cases histories, the paper-based products provide a food source for mold growth, even when the paper is coated with asphalt. The Grade C and particularly the Grade D papers are thin and easily damaged (punctures and tears). The durability of the thinner building papers is of further concern given the many fastener penetrations for attachment of the lath and accessory beads (Figure 3). Many design guides indicate that the more permeable papers (Grade C and D) are desirable to "allow trapped moisture to escape from the wall cavity". We consider this to be poor advice in terms of overall performance of the wall system. The first priority is to keep rainwater from penetrating to the stud wall and, therefore, a highly moisture resistant building paper should be used. While an in-depth discussion of vapor drive in wall systems is beyond the scope of this paper, the Southwest region of the USA and California do not have strong vapor drives either to the interior or exterior. As such, designers should focus on the waterproofing characteristics of the back-up and not its vapor permeability.

Figure 3 – *Accessory Beads Result in a Concentration of Fastener Penetrations*

Installation of the back-up felts in weatherboard or shingle fashion is often cited in design guides, but the end laps between felts can provide open pathways to the stud wall. Sealing end laps as well as sealing all felts to perimeter nailing flanges of windows will increase the reliability of the weather barrier significantly with little additional effort.

Using two layers of building paper (e.g., #15 felt over #30 felt) reaps multiple benefits. Building paper takes abuse from weather and construction prior to stucco application and may lose some of its asphalt content due to leaching as a result of contact with the stucco paste. The outer "sacrificial layer" protects the inner layer from deterioration. Vertical laps should be staggered between layers, thereby reducing the potential for leakage through unsealed or poorly sealed end laps.

We prefer to use a heavy asphalt-saturated felt (#30) back-up with rubberized-asphalt (peel-and-stick) membrane strip flashing at penetrations (e.g., windows) as a reinforcing/sealing strip behind accessory beads that use large fasteners (Figure 4). In some instances, we have noted that stucco formulations with certain admixtures have a tendency to stick to asphalt-saturated felt, reducing the ability for water to drain freely from within the system. Use of kraft paper as a bond breaker can eliminate this problem, but provides a medium that can support mold growth within the wall system.

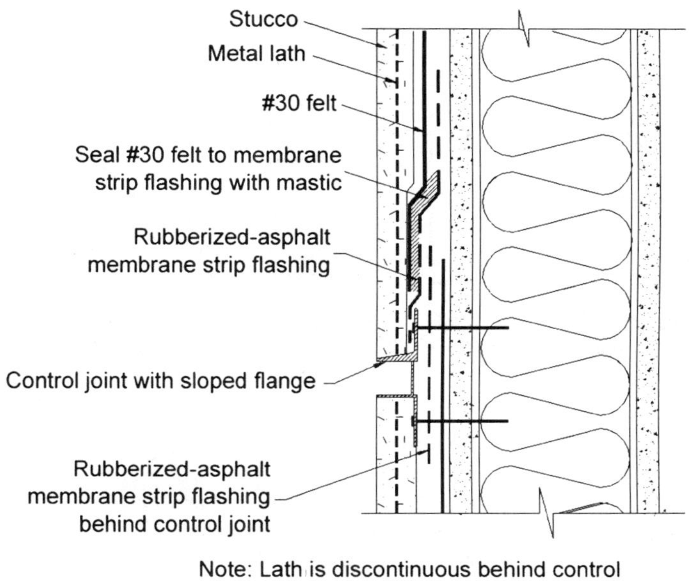

Figure 4 – *Horizontal Control Joint*

Some paper-backed laths are manufactured with a 1-1/2 in. recess of building paper on one edge and one end, with a similar paper overhang on the opposite edge and end. The lath restricts access to end laps for sealing and prevents the code recommended 2 in. horizontal and 6 in. vertical (end) laps.

Special care needs to be taken when installing paper-backed lath to ensure that all laps are "wire-to-wire" and "paper-to-paper"; otherwise, large gaps and pockets are formed when these composite sheets are simply lapped one on top of another. While this may seem obvious, it is a problem that occurs all too often.

Accessories – Weep Screeds, Control Joints, and Casing Beads

A consistent problem that we have found in modern stucco construction is the failure to provide adequate weep capacity for the system, particularly in multi-story construction. Building codes require weep screeds at the base of walls, but not at each floor level. Typically, a control joint is installed at the floor line that consists of two J-shaped screeds separated by a 1/4-in. gap with a solid backing flange (Figure 5a). While these screeds provide crisp edges to the stucco, they do not allow water to weep from the wall. The lack of effective weeps at floor levels results in water accumulating within the system until it finds a defect in the back-up waterproofing and leaks into the building.

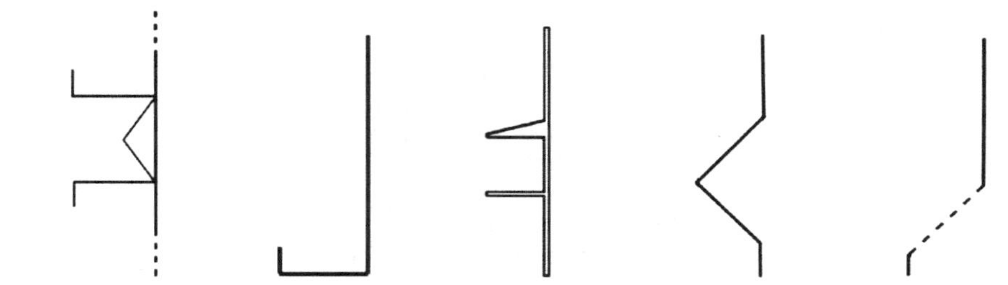

Figure 5 – *Stucco Accessories*

Casing beads (Figure 5b) are typically a solid metal flange without weeps installed at stucco terminations, such as at window perimeters. The beads provide crisp stucco terminations and a surface to which an installer can screed. Casing beads are often installed continuous around window corners by "snipping" the bead flanges, resulting in a flange discontinuity that allows water to leak into the building (Figure 6). Water also bypasses unsealed paper along jambs and flows along jamb casing beads. If building paper installed at the sill does not extend behind jamb casing beads, any water on the jamb beads flows directly behind the building paper.

Figure 6 – *Discontinuity of Casing Bead Flanges at a Window Head Corner*

Weep screeds need to be used frequently and as a general guide should be installed at each floor line in multi-story buildings and at window heads and all other "soffit" returns. Weep screeds that are used at the foundation level typically consist of a solid metal flange that is sloped to the exterior (Figure 5d). A few manufacturers offer a sloped metal flange with holes along the length of the screed that further facilitate drainage (Figure 5e). Similarly, horizontal control joints with sloped flanges are available (Figure 5c).

Figure 4 illustrates a typical installation of a horizontal control joint with a sloped flange, which promotes drainage. The addition of "self-healing" rubberized-asphalt strip flashing behind the control joint accommodates the additional fasteners through the #30 felt. Since the flange is sloped it often lacks factory weeps, which can be drilled prior to installation to further promote drainage. Returning the lath along the sloped flange can prevent stucco from clogging weeps. Control joints with sloped flanges are preferable to those with square flanges, which have short, upturned legs at the exterior. The upturned legs tend to trap water, even in the case of weeped control joints.

Control joints often provide avenues for water penetration to the back-up and act as conduits that collect and carry water to their end terminations. Splice joints, T-intersections between vertical and horizontal beads, and the terminal ends of beads need to be sealed to minimize infiltration through the gaps that occur in the surface barrier at these locations. Also, poor consolidation of the stucco against the edges of the accessories can result in oversized cracks (gaps) when the stucco dries and shrinks away from the control joints.

When constructing control joints (to control the location of cracking) the lath should be terminated at the edges of the accessory and the two elements should be wired together such that the metal lath and flange of the accessory form a smooth planar transition. This helps to maintain a consistent thickness to the stucco and allows the control joint to "float" a bit as the stucco dries and shrinks. The waterproofing membrane should always be continuous behind control joints. In some cases, we find that the accessory is applied on top of a continuous installation of metal lath and that the accessory is screwed in place. The resulting continuity of the lath between panels reduces the probability that cracks will be confined to the edges of the joints and may promote cracking within the field of the panel.

Crack Control

Aside from cracking due to thermal and moisture induced movement, stucco is also prone to cracking at re-entrant corners, changes in thickness, or changes in substrates. Strategic placement of control joints can reduce stucco restraint and control cracking.

Although there are no set standards for spacing of control joints to accommodate drying shrinkage and cyclical thermal movement, it is generally accepted that control joints should be installed to produce panels ranging from 100 to 144 sq ft, preferably in a square shape, and spaced not more than 10 to 12 ft (conservative) in any one direction. Control joints should also be installed over any existing building expansion joints.

We have found that a number of modern stucco buildings exhibit an inordinate number of cracks, even though the stucco is properly panelized by control joints. As discussed above, improper installation of control joints can promote cracking of the stucco. Installation over open framing can also result in cracking problems. Without sheathing, the lath spans between studs and can bulge into stud cavities upon stucco application. This results in a thicker stucco section between the studs, a non-planar sag in the reinforcing, and a thinner section at points of lath attachment over the studs. The combination of these factors invites shrinkage cracks at the stud locations. In general, the thickness of stucco should be monitored during any application to ensure uniformity and avoid "stress concentrations" and resultant cracking at changes in thickness.

Curing

In many instances, improper curing, or a lack of curing, is the source of cracking. Stucco shrinks as it dries, inducing tensile stresses in the stucco that lead to cracking. Similar to concrete, curing of stucco is required to achieve proper hydration of the portland cement. Stucco must dry slowly and uniformly to achieve full strength and minimize cracking.

Codes recommend curing the scratch and brown coats for a minimum of 48 hours each. The delay between coat applications allows each to cure independently. The body of the stucco must be allowed to obtain its initial shrinkage, which usually equates to an interval of 7 days between finish and brown coat applications. The finish coat is typically not cured, as non-uniform wetting and drying can result in color differences within the finish. The code recommendations for curing should be used as general guidelines, and curing times need to be project and site specific. Factors such as temperature, relative humidity, exposure, wind, etc. all need to be considered when determining the length of cure time and time between coat applications. High relative humidity (over 75%) can reduce the frequency of moistening, while heat and wind require increased moistening.

Curing can be achieved through fog-spraying the stucco at intervals throughout the day and early evening to maintain moisture. A fine mist is used to prevent erosion of the stucco surface. Some industry references recommend fog spray twice a day in the morning and evening, but as mentioned previously, curing is site- and project-specific. Vapor barriers such as polyethylene membranes can be installed over the stucco to retain moisture. Membranes mask the stucco and therefore must be monitored for breaches to ensure uniform moisture throughout the veneer.

On multi-story buildings, conditions on the south and west elevations can be extreme in the summertime and may require greater effort to obtain proper cure and preclude rapid shrinkage. This may involve "tarping" the scaffold to reduce exposure to sun and wind as well as monitoring the moisture of the stucco. Scaffolding profiles cast shadows on building walls and can result in areas of differential drying.

In any event, the stucco needs to remain moist during the curing process and we suspect that increasing pressure on time and money in the construction process has led to ineffective control of curing and shrinkage on many projects. Without rigorous monitoring of curing times and procedures, we have found that even the minimum code requirements are not implemented by contractors.

Case Histories – Lessons Learned

High-Rise Building

Some of the current construction practices that have led to problems in the performance of traditional stucco cladding can be well illustrated by an 11-story stucco-clad building that we recently investigated in California. The cladding consists of stucco on self-furring, paper-backed wire lath spanning between steel studs with no exterior sheathing (Figure 7).

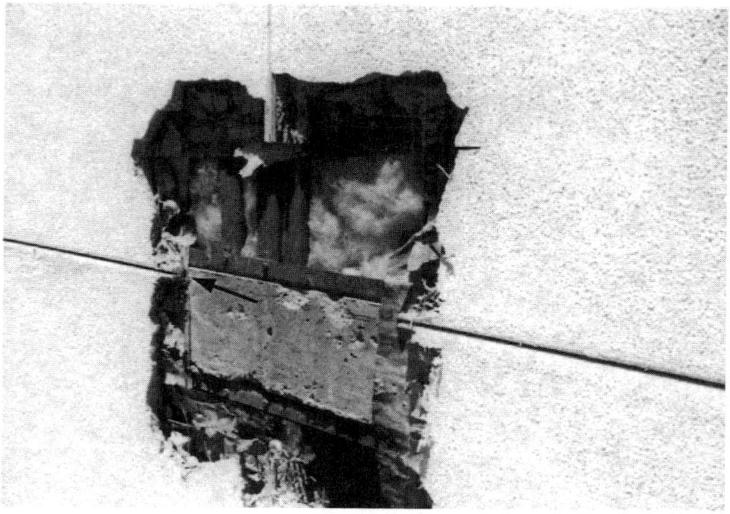

Figure 7 – *Control Joints Compress Paper Against the Slab and Stud Track, Creating a Dam*

The stucco is unsupported between steel studs and tends to bulge into the spaces between them. The lack of sheathing resulted in thicker stucco sections between studs and thinner sections over the studs, and contributed to many areas of unembedded wire lath, often corroded. For the reasons discussed above, vertical cracks developed in the stucco at many stud locations. Stucco cracking is widespread, even though the stucco is properly panelized with control joints. The number of cracks increases on the west and south elevations of the building, which are subject to significant solar exposure and have an open fetch to the prevailing winds.

An interesting effect of the open framing is that water flows relatively freely (promptly) on the building paper. The building paper is not compressed against a sheathing board and the void spaces for drainage are increased significantly. Unfortunately, this had negative consequences for the wall waterproofing because of the defects in the building paper and the installation of control joints. The horizontal control joints used at each floor line are the solid J-shaped screeds that preclude effective weeping (drainage to the exterior). Also, the horizontal control joints are fastened to the stud tracks and outside face of the concrete floor slabs, which compresses the back-up assembly and creates a dam at each floor line (Figure 7). As a result, water that drains downward on the building paper backs-up at the floor line and either overtops horizontal laps in building paper or leaks through holes, tears, or fastener penetrations (Figure 8). The leakage is severe because water rapidly flows to the compressed areas and forms relatively large reservoirs in the vicinity of the control joints.

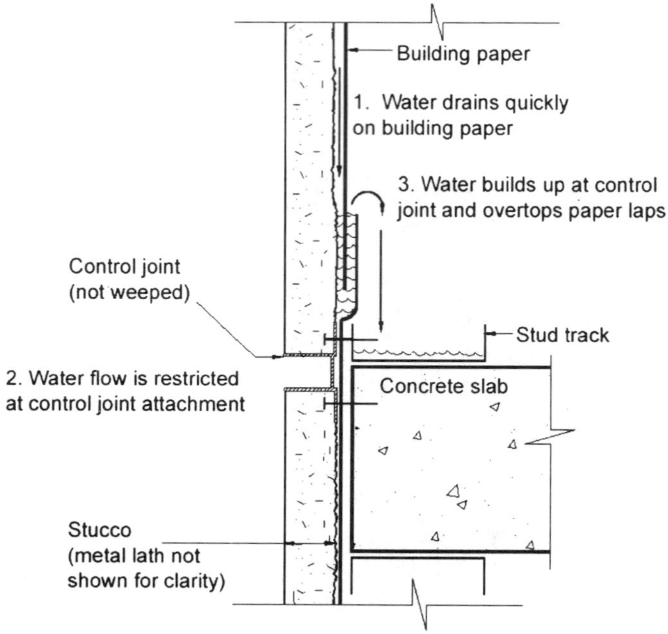

Figure 8 – *Damming Effect at Floor Line*

To make matters worse, the paper-backed lath utilized a Grade D backing. This grade of paper is relatively thin, easily damaged, and will deteriorate when subject to prolonged moisture exposure. In fact, this paper was a medium for mold growth, and our sample openings in the exterior walls showed significant mold growth throughout the building paper including the asphalt-coated surface. We also found many "paper-to-wire" laps in the lath that proved to be significant leakage pathways during our water tests.

It is interesting to note that the governing building code only requires a weep screed at the base of the walls and allows the use of Grade D paper on a high-rise building. Installation errors aside, the use of a thin, paper-based back-up with no means to drain water to the exterior for most of the wall height seems questionable at best. Accordingly, we think that code officials may want to examine the fundamental requirements for stucco used in high-rise construction.

To control re-entrant corner cracking, vertical control joints were installed coincident with window jambs. The vertical joints deposit water to window heads and contribute to leakage around windows. Although the true cause of leakage is the lack of head flashing or reliable back-up waterproofing, designers need to be aware of the risks associated with control joints that terminate at wall penetrations. Figure 9 shows a continuous head flashing turning down over a window. The flashing has no transverse joints and corners are soldered (or welded) watertight.

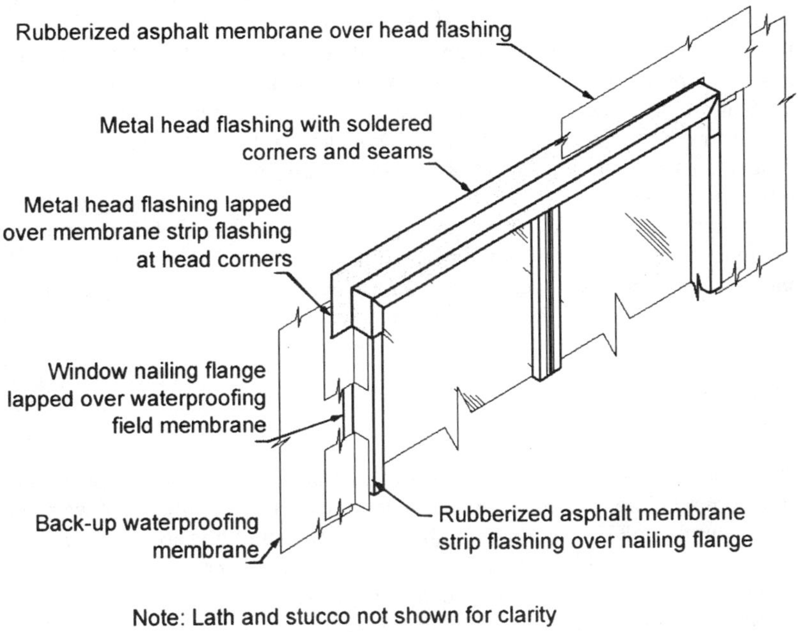

Figure 9 – *Figure Flashing at Window Head*

Mid-Rise Building

Our next case history involves a large complex located in Southern California that comprises City Hall offices and a Civic Center. The wall construction consists of stucco applied to self-furring metal lath, building felts (one or more layers of asphalt-saturated felts), gypsum board sheathing, and steel studs. Although the leakage problems are not as pervasive as in our hi-rise case history, they are significant and occur throughout the complex. In addition, the stucco uses a polymer modifier and an integral color finish coat that exhibits widespread cracking and discoloration.

From a waterproofing standpoint, the fundamental problem is a lack of weeps. All of the horizontal control joints and accessories at window soffits reveals are solid screeds. Another problem with the accessories is that splices between pieces and the terminal ends are unsealed and provide large gaps for water to penetrate to the paper and eventually to the sheathing through fastener holes and dry end laps in the felts.

Another common problem is the design of nearly flat window shelves within the facade that are clad with stucco (i.e., the use of stucco as a roof). In some cases these shelves are nearly 2-ft deep and are deteriorating rapidly due to leakage, even though they have a back-up peel-and-stick membrane underlayment. The water ponds on the membrane and eventually leaks through fastener holes.

As to the problems involving polymer additives and integral color finish coats, suffice to say that proprietary mixes introduce additional variables that must be controlled during application. Designers are well advised to research past performance of all such products. Lower-cost polymers such as styrene butadiene rubber (SBR) are less resistant to ultraviolet than acrylic polymers and admixtures for air-entrainment may cause the stucco to bond to asphalt papers. Reasonable panel sizes and proper panel geometry that eliminate reentrant corners are always a starting point for crack control, and curing procedures must be monitored for initial results (effectiveness) and changed if necessary for the balance of the project. Once again, it appears on larger buildings that additional measures are needed to protect the stucco during cure on the south and west elevations.

Low-Rise Buildings

Problems in low-rise and residential construction are fairly commonplace and often result from a failure to execute simple, repetitive details. To illustrate, we recently investigated a number of 3- to 4-story wood-framed, multi-building apartment complexes in Texas. Each of these buildings has stucco on unfurred diamond metal lath and asphalt-saturated felt paper over oriented strand board (OSB) sheathing.

Although the code allows its use, unfurred lath resulted in many areas of unembedded lath. The stucco compresses the lath against the building paper and sheathing, resulting in an imprint of the diamond mesh on the paper. The building paper tears along the imprinted lines of the mesh, creating discontinuities that result in leakage to the building interior.

As is often the case, the majority of the leakage problems at these buildings result from improper integration of stucco with other building components, such as windows. The windows are aluminum with integral nailing fins installed within punched openings. Although reference guides clearly show how to shingle building paper to the window nailing fins, almost invariably they are mis-shingled during construction (Figure 10). A peel-and-stick membrane was installed to strip the nailing flanges to the sheathing. The building paper was then installed over the sheathing, resulting in a correct overlap at the window head and an incorrect lap at the windowsill. Another problem we often find is that the building paper is set dry (i.e., without a seal or mastic) along the window jambs, which allows water to wrap the edge of the paper to unprotected sheathing.

Figure 10 – *Building Paper is Often Mis-shingled to Window Nailing Flanges*

It is interesting to note that at each of these buildings, window perimeters have remnants of protective polyethylene sheets that were installed over the windows prior to stucco installation. It is common for contractors to install polyethylene over the windows to keep them clean during stucco installation, and to cut the polyethylene away after stucco cures. This results in a continuous strip of polyethylene at the window perimeter. At the sill, the polyethylene is compressed against the window frame and laps behind the already mis-shingled building paper (Figure 11).

The polyethylene strip contributes to building leakage and sheathing deterioration. At each of these buildings, the jamb extrusion of one window is dry-set into that of the adjacent window, forming a common "ganged" window (Figure 11). The ganged mullion creates a race between windows in which water collects. Water within the race flows to the sill and is deposited to the backside of the polyethylene strip. The polyethylene directs the water behind the building paper to unprotected sheathing and has resulted in sheathing deterioration.

Figure 11 – *Remnant of Polyethylene Installed Over Windows Prior to Stucco Application*

Poor integration of stucco with roofing is also a major source of leakage at some of these buildings, namely at terminations of roof rakes within fields of stucco walls. Rake flashings lack metal diverters or "kick-outs" at their terminations, and building paper does not extend up and behind the rake flashing (Figure 12). As a result, the flashings deposit large amounts of roof drainage water directly behind the building paper, which has led to deterioration of 3 stories of OSB sheathing beneath the rake terminations. A metal diverter should be installed at rake terminations, and care needs to be taken to properly integrate the building felts and roof underlayment with the rake flashing as shown in Figure 13.

Figure 12 – *Lack of Metal Diverter, and Building Paper Does Not Extend Behind Flashing*

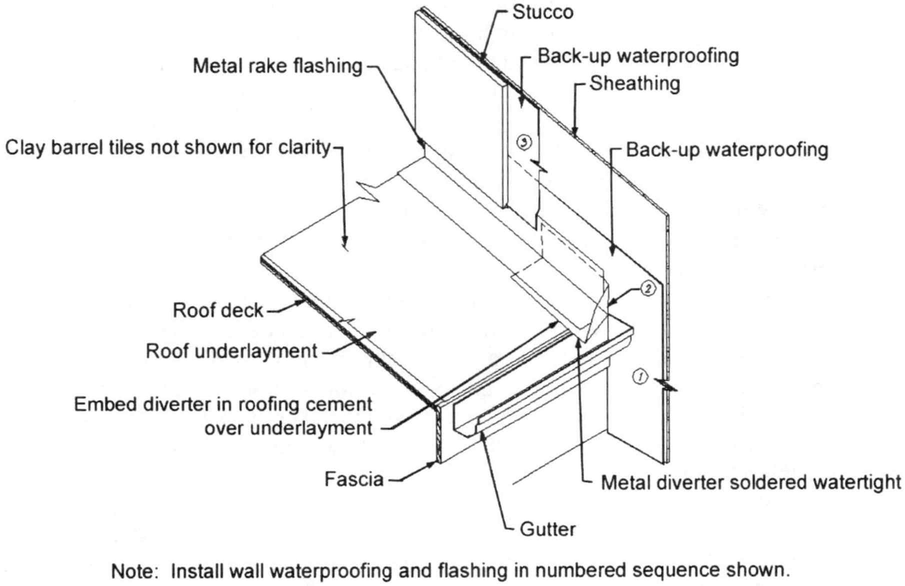

Figure 13 – *Integration of Stucco with Roof Flashing*

Summary Recommendations

As a result of our findings in recent investigations, we recommend that the industry (designers, contractors and code officials) consider the following recommendations to improve the performance of stucco cladding systems:

- Install durable exterior sheathing over stud walls. Sheathing helps to control stucco thickness, which in-turn minimizes cracking, facilitates embedment of lath, and regulates the flow of water over building paper. The risks associated with not using an exterior sheathing are not worth the cost savings.
- Consider the use of a peel-and-stick membrane for the back-up waterproofing layer. As a minimum, use a heavy grade asphalt-saturated felt with sealed end laps and peel-and-stick membrane around wall penetrations and at areas where large or numerous fastener penetrations are expected, such as behind accessories. The peel-and-stick membranes have the advantage of sealing themselves to fastener penetrations, and all laps are fully sealed. The peel-and-stick membranes are not as vapor permeable as the felts, and a moisture drive analysis should be performed for special occupancies that generate high interior moisture levels (e.g., a swimming pool facility or computer room). However, for most occupancies in

California and the Southwest, vapor drive and condensation within the wall are not at issue – keeping rainwater out of the wall is paramount.

- Consider using two waterproofing layers rather than one. Use the traditional method of installing paper and lath in separate applications rather than using paper-backed lath. The paper-backed laths hinder or preclude sealing of end laps and are sometimes supplied with an integral kraft paper between lath and building paper which serves as a medium that supports mold growth. Also, paper-backed laths all too often lead to wire-to-paper laps.
- Properly fur-expanded metal lath to promote embedment and avoid tearing along the imprinted lines that can result in the building paper when the lath is placed tightly to the back-up.
- Provide weeps at all floor lines (regardless of building height) and soffit returns. Consider using a two-piece, weeped control joint at floor lines. Inadequate drainage of the back-up system is a consistent problem in multi-story buildings.
- Place control joints in strategic locations to control cracking, but realize that control joints provide paths for water to bypass the exterior skin. Terminations and splices should be sealed. While we recommend watertight flashings at all wall penetrations, they are particularly essential where control joints terminate at penetrations.
- Although codes stress curing and delineate reasonable minimum requirements, designers and installers need to monitor the process to ensure that adequate cure is achieved and that environmental factors that promote rapid shrinkage are controlled.

References

[1] "Standard Specification for Application of Portland Cement-Based Plaster" (ASTM C926-98a).

[2] "Guide to Portland Cement Plastering" ACI 524R-93, American Concrete Institute.

[3] Isberner, A. W., Jr., and Melander, J. M., "Portland Cement Plaster (Stucco) Manual," Portland Cement Association, 1980.

[4] Gorman, J. R., Jaffe, S., Pruter, W. F., and Rose, J. J., "Plaster and Drywall Systems Manual," 3rd ed., McGraw Hill, 1988.

Edward S. Lindow,[1] *and Lee F. Jasinski*[2]

Panelized Wall Construction: Design, Testing, and Construction Procedures

---

Reference: Lindow, E. S., and Jasinski, L. F., **"Panelized Wall Construction: Design, Testing, and Construction Procedures,"** *Performance of Exterior Building Walls, ASTM STP 1422*, P. G. Johnson, Ed., ASTM International, West Conshohocken, PA, 2003.

Abstract: Today's building construction requires speed, efficiency, and economy. To meet these needs, prefabricated wall panels are being incorporated in curtain wall systems, creating a rapidly expanding market. To illustrate concepts, benefits and needed research and development, the design and use of a panelized wall system are discussed. The prefabricated panels are manufactured using steel studs and a mechanically fastened rigid board sheathing. The exterior envelope can then be completed using a factory installed EIFS (exterior insulation finish system) or a conventionally built brick masonry veneer. The development and constructability of the panelized curtain wall system are described along with associated design and code validation testing. In addition, the need to integrate water management details into the curtain wall design is also reviewed. A current construction project provides the transition from design to installation and exemplifies the benefits of panelized construction.

Keywords: exterior wall panels, curtain walls, EIFS

Introduction

In today's fast track, target-priced construction environment, innovation is a necessity to remain competitive. This is a fact for the design professional as well as for the general and specialty contractors. The key to success is to provide technical and economic innovation without sacrificing the quality of the end product. New construction materials or techniques, which are not adequately designed and tested only add to the risk of the entire project team.

---

[1]Vice President, Soil and Materials Engineers, Inc. 43980 Plymouth Oaks Blvd., Plymouth, MI 48170.

[2]President, Jasman Truss and Panel Technologies, 1175 E. North Territorial Road, Whitmore Lake, MI 48189.

The manufacturing industry has demonstrated the benefits of prefabricating components into subsystems. For example, an automotive assembly line now consists of combining prefabricated subsystems such as doors, dashboards and even engines which are manufactured off site by tiered suppliers. Speed, economy and quality are the goals of this delivery technique.

In the construction industry, prefabrication of exterior wall panels is not a new strategy. Precast concrete panels and a variety of curtain wall systems have been used over the last 30 or so years. However, as experience is gained and performance documentation becomes available, the use of exterior wall panels continues to evolve. This paper discusses several aspects of the design, testing, and construction of a prefabricated exterior wall panel system.

Panel Design

The exterior wall panel assembly (EWPA) discussed in this paper is a prefabricated steel stud frame with exterior sheathing and anchorage components attached at the factory (Figure 1). The steel studs are 6 inch, 16 gauge, and spaced at 16″ o.c. The head and sill tracks are 14 gauge. A ½″ Dens-Glass[R] sheathing is attached to the studs with mechanical screw fasteners. This type of sheathing combines embedded glass mats with a water resistant treated gypsum core to provide a durable substrate for the selected façade. The panel can be designed for inclusion in a brick veneer curtain wall system or with an Exterior Insulation and Finish System (EIFS).

Figure 1 – *Basic wall panel fabrication.*

The brick veneer panel (Figure 2) includes a vapor retarder, brick ledge, brick ties, and cavity flashings installed on the steel stud frame. The masonry veneer is then installed after the panel is secured in place. This assembly expedites the closing in of the building so other trades can continue construction. It also reduces coordination conflicts concerning installation of the shelf angle and flashings and promotes continuity of the water management system.

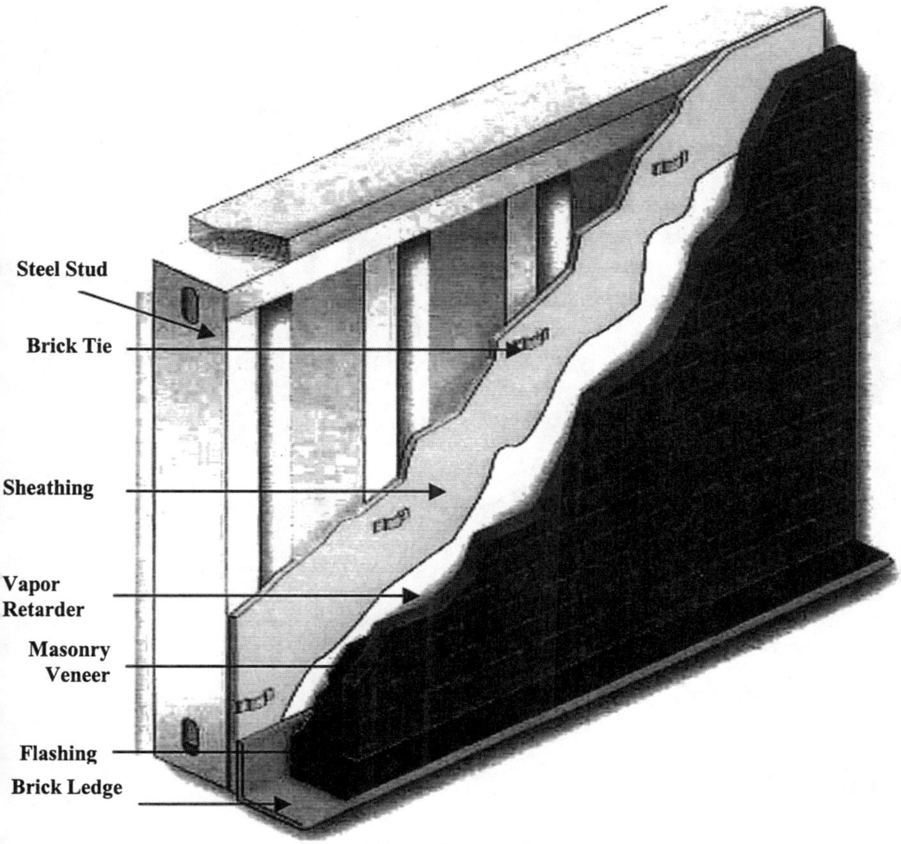

Figure 2 – *Brick veneer panel.*

The EIFS panel is preassembled with the insulation board and lamina applied to the sheathing of the basic steel stud frame (Figure 3). Prior to application of the EIFS, the joints in the sheathing are taped and a waterproof coating is applied. The EIFS insulation board is applied in vertical ribbons of adhesive and a starter track with weep holes is provided to evacuate entrapped moisture.

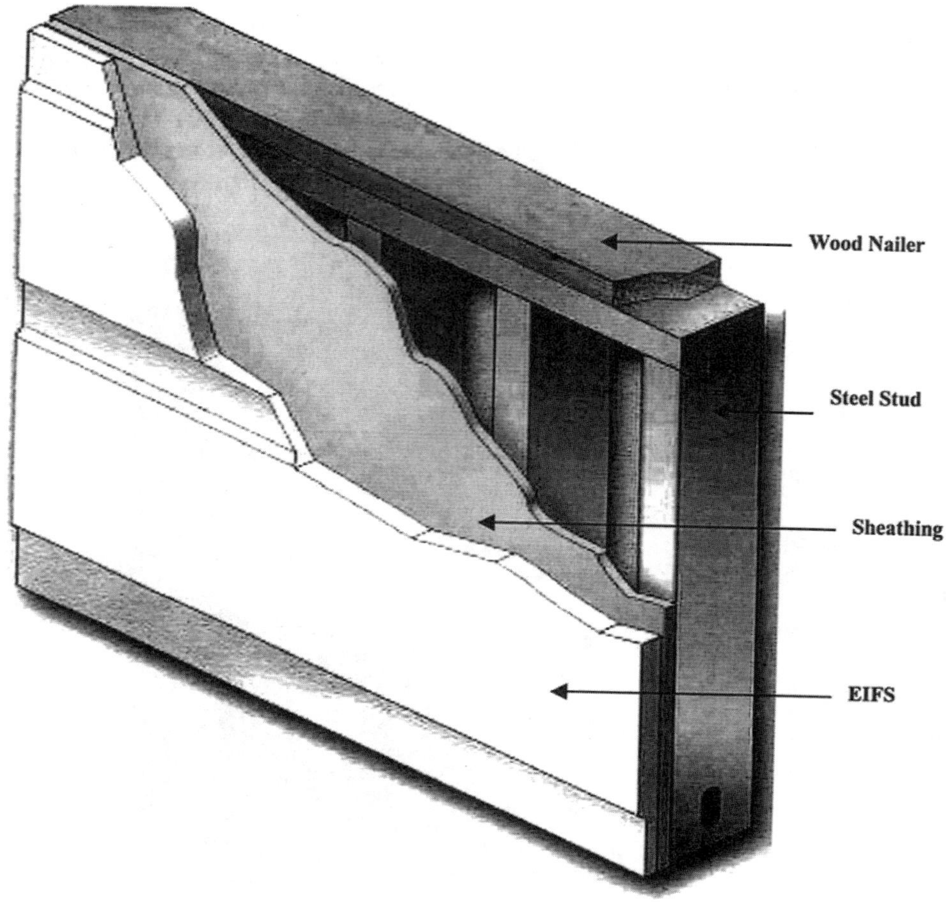

Figure 3 – *EIFS Panel.*

Testing Criteria

While there are numerous standard test methods for component wall materials, procedures to assess the wall system performance are relatively sparse. Table 1 lists standard tests typically associated with EIFS wall panel assemblies. In addition, masonry veneers, steel components, and fenestration elements all have standard tests to characterize individual materials used in the construction of exterior walls. However, there is no test method or code that can be used to assess the wall panel performance as an as-built system.

Table 1 – *Typical Test Criteria*

Test	Method	Criteria
Abrasion Resistance	ASTM D968	500 liters of sand
Accelerated Weathering	ASTM G23	2,000 hrs.
Accelerated Weathering	ASTM G53	200 hrs.
Adhesion	ASTM C297	>15 psi
	After accelerated weathering and freeze thaw	>5psi
Freeze Thaw Resistance	ASTM C67	60 cycles
Mildew Resistance	ASTM D3273	28 days
Salt Spray	ASTM B117	300 hrs.
Water Penetration	ASTM E331	6.24 psf for 15 min.
Water Resistance	ASTM D2247	14 days
Wind Load	ASTM E330	Per Code
Surface Burning	ASTM E84	Flame spread <25 Smoke dev. <450
Full Scale Div. Fire Test	ASTM E108	No excess flame spread.

Full Scale Panel Testing and Analysis

The structural performance of the standard panel assembly was tested using Standard Test Method for Structural Performance of Exterior Windows, Curtain Walls, and Doors by Uniform Static Air Pressure Difference (ASTM E330). The test panel was nominally 10′ x 20′ and included the steel stud framing and Dens-Glass Gold Sheathing. Figures 4 and 5 show the back of the test panel and front of the test chamber. The test panel remained fully functional during the testing sequence. Based on test results, the wall panel assembly meets acceptance criteria for 90 mph wind load for Commercial Construction with mean roof heights of 100′ (Exposure B) or less for a design load of 20.5 psf windward, and for an 80 mph wind load for Exposure C with a mean roof height of 50′ or less. In addition, the panel meets acceptance criteria for 90 mph wind load with mean roof heights of 20′ for design loads of 22.3 psf leeward, and for an 80 mph wind load for Exposure C with a mean roof height of 40′.

The prefabricated panel assembly was also analyzed in accordance with BOCA Chapter 16, Structural Loads. A linear elastic finite element analysis was performed on the standard panel with a 2″ cavity and brick veneer. Using a wind speed of 70 mph and a building height of 30′ the maximum wind pressure on the panels is 18 psf. When the sheathing is attached with three fasteners per stud and brick ties are spaced 16″ horizontally and 24″ vertically, the required pull out resistance of fasteners and ties is 40 lbs. to meet the wind load criteria.

Figure 4 – *Back of test panel.*

Figure 5 – *Front view of test chamber.*

Construction Example

This exterior wall panel assembly has been used on numerous projects over the past three years. The experience gained with each project has produced improvements in the design, manufacturing and construction of the panels. Feedback from erection crews has been especially useful in developing connection and positioning techniques.

A contemporary example is a four story, 180,000 square foot office building being built in Auburn Hills, Michigan. Construction was begun in late 2000 with the exterior wall scheduled for completion during the winter of 2001. The building consists of three wings (each with 48,000 s.f. of floor space) radiating from a central atrium area. The curtain wall system includes horizontal bands of strip windows and EIFS accent panels. The EIFS panels are 30′ in length and 8′8″ in height.

Some of the innovations and lessons learned on this project are summarized below.

- A constructability review was used to integrate the various curtain wall components into the building envelope construction. Constructability review topics are discussed in the next section.
- The EIFS wall panels were installed prior to placing the concrete floors to expedite the construction schedule. The window openings were closed in with plastic sheet material allowing the area to be heated and the concrete floors installed (see Figure 6). The time savings were significant. On one wing, the panels were installed and the building enclosed in just eleven days after the steel erection was complete.
- The weight of the panel was carried by clip angles at each column of the 30′ bays (see Figure 7). The panels were secured by welding the metal track to the clip angle. This connection carried the dead weight of the panel.
- Specially fabricated clips (see Figure 8) were designed for temporary resistance of wind loads and to maintain the panel plumb. The clip is welded to the floor edge angle to provide wind resistance and the panel is vertically aligned using the slotted connection.
- Following floor placement, the horizontal and vertical alignment of each panel was readjusted to compensate for movements caused by the additional load of the concrete floors. In addition, a kicker (brace) is installed from the floor to the bottom of the panel to complete the structural attachment.
- The wall panel subcontractor retained a third party agency to provide quality control during fabrication and erection and to review the integration of the wall panel system with other components of the building envelope. The latter is a necessity since the integration normally occurs after the curtain wall contractor has left the site. Quality assurance must be performed to verify continuity and coordination of roof flashings, window flashings, and sealant application.
- Some of the water management details included: double sealant assemblies at window heads, a metal sill pan with end dams at sills, vertical panel joints employed a standard sealant assembly (ie, backer road and bead of sealant), and a

vapor retarder was added to the inside of the wall framing prior to installation of drywall.

Figure 6 – *Panel erection.*

Figure 7 – *Panel connection at column.*

Figure 8 – *Clip connection to floor angle.*

Constructability Review

In simplistic terms, the architectural drawings for a project provide the material concept, geometry, and appearance required for the building's exterior façade. Shop drawings by individual subcontractors supply details on the materials and methods proposed to satisfy the design intent. A constructability review should be designed to assess compliance with project requirements, coordinate various subsystem and building components, and reduce field requests for information. It is not value engineering nor economic streamlining (ie, cost reduction) of the construction.

For exterior wall panel assemblies, the constructability review should include the following items.

- Compliance with specifications and drawings
- Appropriateness and compatibility of materials
- Areas of coordination between trades or subs
- Throughwall flashings at head, sill, and other openings
- End dams at termination of all flashings
- Effective weeps to the exterior
- Identify thermal bridges or areas of potential condensation
- Prefabricate flashings for wall penetrations (utilities, vents, etc.)
- Continuity of water management between panels
- Integrity of roof to wall detail

- Mechanical fasteners puncturing flashing
- Weeps located behind sealant
- Redundancy of sealant joints
- Assess maintainability of sealants
- Effect of dual vapor retarder or air barriers if present
- Anticipated interior temperature/humidity levels
- Accommodation of thermally induced movement
- Effect of changes in insulation thickness

In addition, the submittals for every project should include: temperature gradient through the wall, vapor pressure gradient through the wall, analysis of condensation potential, and a full scale mock-up.

Exterior Wall System Maintenance

Just as for HVAC and elevator equipment, each building project should include a Maintenance Manual for the wall system to be provided to the owner at project completion. In addition to specific warranties and contacts, the manual should detail inspection frequencies and maintenance methods. It should also clearly communicate to the owner that these systems will require maintenance at periodic intervals during their service life. For example, building sealants should be programmed for maintenance or replacement on a 5 to 7 year basis. An EIFS system should be recoated on a 10 to 15 year cycle. Weep systems for masonry veneer and window systems should be inspected at least annually. This type of information will be invaluable to the facility manger and should also reduce conflicts over performance and extend the serviceable life of the building envelope.

Conclusions

A curtain wall has three primary functions: keep water out of the occupied space, resist wind pressure and suction, and look good. Naturally, there are numerous other characteristics involved in the selection of the curtain wall such as cost, construction, schedule, thermal efficiency, level of maintenance required, and structural load considerations. With proper design and quality installation, panelized wall construction, such as described herein, can effectively meet these functions and provide value to the building project.

Based on current experience with the exterior wall panel assembly, the following benefits can be realized.

- Reduce the effect weather has on the construction schedule
- Improve the uniformity of wall construction

- Increase the quality of wall components by construction in a controlled environment
- Facilitate quality control activities during fabrication
- Exterior wall erection time can be shortened by about 75%
- Staging area required is reduced since wall materials are generally not stored on site
- Just-in-time delivery also reduces storage and traffic conflicts
- Interior finishes can commence as soon as the building is enclosed since the steel studs provide an excellent foundation for interior walls
- Scaffolding is not necessary when complete panels can be erected by crane
- Economy to the project and improved cost control will be produced.

Working with the panelized wall systems has also generated areas of needed research or development. The following items require further study to advance the state of the art for panelized wall construction.

- Performance criteria for this type of wall system must be developed or refined. Compliance with building codes alone does not address quantitative assessment of the wall serviceability.
- The effectiveness of internal EIFS drainage features should be verified, both as built and with time.
- Continuity of thermal and moisture control across panel joints should be evaluated.
- Service life and maintenance history data should be developed for prefabricated wall panels as well as built-in-place wall systems.
- Life-cycle cost models should be developed for wall panel systems.

Richard A. Behr[1] and Heinrich Wulfert[2]

A Wall System that Inherently Satisfies Proposed NEHRP Seismic Design Provisions for Architectural Glass

---

Reference: Behr, R. A., Wulfert, H., **"A Wall System that Inherently Satisfies Proposed NEHRP Seismic Design Provisions for Architectural Glass,"** *Performance of Exterior Building Walls, ASTM STP 1422,* P. G. Johnson, Ed., ASTM International, West Conshohocken, PA, 2003.

Abstract: Proposed Year 2000 revisions to the 1997 *NEHRP Recommended Provisions for the Seismic Regulations for New Buildings and Other Structures* (FEMA 302) include first-generation seismic design provisions written specifically for architectural glass components. Scheduled for publication by the Federal Emergency Management Agency (FEMA) as FEMA 368 in 2001, the 2000 updates of the *NEHRP Provisions* will elevate the degree of design attention paid to seismic life safety issues associated with architectural glass components in exterior building wall systems.

In an independent effort, an "Earthquake-Isolated Curtain Wall System" (EICWS) has been developed and tested in the Building Envelope Research Laboratory at Penn State University. By decoupling each story level of the wall system from adjacent story levels, the EICWS has shown an inherent ability to accommodate large interstory displacements in the vertical, horizontal and out-of-plane directions without jeopardizing life safety or compromising wall system serviceability. Thus, the EICWS has demonstrated an inherent ability to satisfy and exceed, by a wide margin, the proposed seismic design provisions for architectural glass that are now being considered for adoption in the 2000 *NEHRP Provisions*. This paper will include a description of the newly proposed *NEHRP Provisions* for architectural glass and a summary of recent laboratory tests of the Earthquake-Isolated Curtain Wall System.

Keywords: architectural glass, earthquake, seismic design, curtain wall, interstory drift, seismic decoupling

---

[1] Professor and Head, Architectural Engineering Department, The Pennsylvania State University, 104 Engineering Unit A, University Park, PA 16802.

[2] President, Alumino Constructa, Av. Ninos Heroes 2485, Guadalajara, Mexico, C.P. 44520.

Introduction

Post-earthquake surveys have indicated that architectural glass is susceptible to damage resulting from earthquake-induced drifts in the building frame [1, 2]. Provisions for the seismic design of architectural glass in building codes have historically been non-existent or limited to a general statement prescribing, in essence, that "drift be accommodated" [3]. No distinctions have been made in codes regarding the seismic performance of different glass types, different wall frames, and different glazing types. Yet, significant differences exist in the performance of various dry glazed glass types subjected to dynamic, in-plane (horizontal) racking displacements tested in laboratory conditions intended to simulate interstory drifts during an earthquake [4, 5].

The combination of potential life safety risk during earthquakes and the absence of specific seismic design provisions for architectural glass in U.S. model building codes led the first author to propose seismic design provisions for architectural glass for inclusion in the Year 2000 update of the *1997 NEHRP* (National Earthquake Hazards Reduction Program) *Recommended Provisions for Seismic Regulations for New Buildings and Other Structures* [6]. These seismic design provisions were adopted by consensus of Technical Subcommittee 8 on Mechanical/Electrical Systems and Building Equipment and Architectural Elements of the *2000 NEHRP Provisions* Update Committee in 1999. The *2000 NEHRP Provisions* are scheduled to be issued by the Federal Emergency Management Agency as FEMA 368 in 2001. Historically, *NEHRP Provisions* are considered for adoption as model building code provisions in subsequent code editions. Thus, it is likely that the *2000 NEHRP Provisions* for the seismic design of architectural glass will be considered for adoption in the 2003 edition of the International Building Code (IBC).

In a separate, yet related, effort Wulfert and Behr [7] devised an "Earthquake-Immune Curtain Wall System" to increase the serviceability and life safety performance of curtain wall systems under earthquake loads. They claim that this system, which can be adapted to stick-built, panelized, and other curtain wall frame types, is essentially "immune" to damage resulting from swaying motions in the building frame. For a curtain wall system to be "immune" to earthquake-induced damage, the authors submit that it must: (1) show no signs of serviceability degradation (e.g., frame distortion, glass cracking, weather seal damage, increased air or moisture infiltration, etc.) during a moderate earthquake; and (2) show no signs of exceeding an ultimate limit state (e.g., glass fallout) during a severe earthquake. This system is called the "Earthquake-*Isolated* Curtain Wall System" (EICWS) in this paper.

An experimental study was conducted on an EICWS mock-up to verify that the structural isolation incorporated within the EICWS concept imparts an "immunity" to earthquake-induced damage. The research hypotheses were that no glass cracking or glass fallout would occur in the EICWS during simulated interstory drifts, and that no signs of serviceability degradation would occur up to a drift index of 2%, which is representative of interstory drift limits for life safety in model building codes. Results of the laboratory study performed on the EICWS mock-up to test the research hypotheses are summarized in this paper. Comparisons are also made between the response of the EICWS to that of a comparable conventional curtain wall system tested under similar dynamic displacement conditions.

The scope of this paper includes a presentation of the proposed 2000 NEHRP seismic design provisions for architectural glass and an introduction to the "Earthquake-Isolated Curtain Wall System" (EICWS) that inherently satisfies these new seismic design provisions for architectural glass.

NEHRP Seismic Design Provisions for Architectural Glass

Performance of architectural glass in earthquakes can fall into one of four categories:

(a) The glass remains unbroken in its frame or anchorage.

(b) The glass cracks but remains in its frame or anchorage while continuing to provide a weather barrier, and be otherwise serviceable.

(c) The glass shatters but remains in its frame or anchorage in a precarious condition, liable to fall out at any time.

(d) The glass falls out of its frame or anchorage, either in fragments, shards, or whole panels.

Categories (a) and (b) provide both life safety and immediate occupancy levels of performance. In the case of category (b), even though the glass is cracked, it continues to provide a weather enclosure and barrier, and its replacement can be planned over a period of time. (Such glass replacement need not be performed in the immediate aftermath of the earthquake.) Categories (c) and (d) cannot provide for immediate occupancy, and their provision of a life safety level of performance depends on the post-breakage characteristics of the glass and the height from which it can fall. Tempered glass shatters into multiple, pebble-size fragments that fall from the frame or anchorage in clusters. These broken glass clusters are relatively harmless to humans when they fall from limited heights, but when they fall from greater heights they could be harmful.

Included below are the verbatim seismic design provisions for architectural glass included in the *2000 NEHRP Provisions* (FEMA 368).

Actual section, equation, table and reference numbers from the proposed *2000 NEHRP Provisions* are cited below for accuracy and ease in referencing the *Provisions*.

"**6.2.10.1 General -** Glass in *glazed curtain walls, glazed storefront wall systems* and *glazed partitions* shall meet the relative displacement requirement of Eq. 6.2.10.1-1:

$$\Delta_{fallout} \geq 1.25\, ID_p \text{ or } 0.5 \text{ in. (13mm), whichever is greater.} \qquad (6.2.10.1\text{-}1)$$

where:

$\Delta_{fallout}$ = the relative seismic displacement (drift) causing glass fallout from the curtain wall, storefront or partition (Section 6.2.10.2)

D_p = the relative seismic displacement that the component must be designed to accommodate (Eq. 6.1.4-1 [from *NEHRP Provisions*]). (D_p shall be determined over the height of the glass component under consideration.) and

I = occupancy importance factor (Table 1.4 [from *NEHRP Provisions*]).

Exceptions

1. Glass with sufficient clearances from its frame such that physical contact between the glass and frame will not occur at the design drift, as demonstrated by Eq. 6.2.10.1-2, shall be exempted from the provisions of Eq. 6.2.10.1-1:

$$D_{clear} \geq 1.25\, D_p \qquad (6.2.10.1\text{-}2)$$

where:

$$D_{clear} = 2c_1 \left(1 + \frac{h_p}{b_p} \cdot \frac{c_2}{c_1}\right)$$

h_p = height of the rectangular glass,
b_p = width of the rectangular glass,
c_1 = clearance (gap) between vertical glass edges and the frame, and
c_2 = clearance (gap) between horizontal glass edges and the frame.

2. Fully tempered monolithic glass in *Seismic Use Groups* I and II located no more than 10 ft (3 m) above a walking surface shall be exempted from the provisions of Eq. 6.2.10.1-1.

3. Annealed and heat-strengthened laminated glass in single thickness with interlayer no less than 0.030 in. (0.76 mm) that is captured mechanically in a wall system glazing pocket, and whose perimeter is secured to the frame by a wet glazed gunable curing elastomeric sealant perimeter bead of 1/2 in. (13 mm) minimum glass contact width, or other approved anchorage system, shall be exempted from the provisions of Eq. 6.2.10.1-1.

6.2.10.2 Seismic Drift Limits for Glass Components: $\Delta_{fallout}$, the drift causing glass fallout from the curtain wall, storefront or partition, shall be determined in accordance with "Ref. 6-19" [8] or by engineering analysis."

Simply stated, Eq. 6.2.10.1-1 requires that the resistance to glass fallout of an individual glass panel be greater than the relative seismic displacement that the component is designed to accommodate. In the absence of special "drift accommodating" connections between the main building frame and the curtain wall framing members, this relative seismic displacement demand is governed by the calculated seismic interstory drifts for the specific building being designed for site-specific earthquake conditions.

Eq. 6.2.10.1-2 is derived from *Earthquake Safety Design of Windows*, published in November 1982 by the Sheet Glass Association of Japan. Eq. 6.2.10.1-2 is derived from a similar equation in Bouwkamp and Meehan [9] that permits calculation of the interstory drift required to cause glass-to-frame contact in a given rectangular window frame. Both equations are based on the principle that a rectangular window frame (specifically, one that is anchored mechanically to adjacent stories of the primary structural system of the building) becomes a parallelogram as a result of interstory drift, and that glass-to-frame contact occurs when the length of the shorter diagonal of the parallelogram is equal to the diagonal of the glass panel itself.

The 1.25 factors in Eqs. 6.2.10.1-1 and 6.2.10.1-2 reflect uncertainties associated with calculated inelastic seismic displacements in building structures. Wright [10] stated that "post-elastic deformations, calculated using the structural analysis process, may well underestimate the actual building deformation by up to 30%. It would therefore be reasonable to require the curtain wall glazing system to withstand 1.25 times the computed maximum interstory displacement to verify adequate performance." Therefore, Wright's comments form the basis for employing the 1.25 factor in Eqs. 6.2.10.1-1 and 6.2.10.1-2.

The "Earthquake-Isolated Curtain Wall System"

A curtain wall system has been developed by Wulfert and Behr [7] to provide high resistance to earthquake-induced building motions. The Earthquake-Isolated Curtain Wall System (EICWS) endeavors to achieve interstory structural isolation by employing a continuous "seismic decoupler joint" at each story level to isolate the vertical mullions at each story level from the vertical mullions at the story above and/or the story below. Since the decoupler joint isolates the vertical mullions at each story, a specialized structural support system is needed to attach the vertical mullions to the building frame at each story level. Consequently, in-plane and out-of-plane movements between adjacent stories in the building frame due to earthquake-induced ground motions should induce no significant forces within the earthquake-isolated curtain wall frame.

Description of the EICWS

Schematic depictions in Figure 1 contrast the fundamental vibration modes of a typical building frame clad with a conventional curtain wall system with that of the same building frame clad with an EICWS. Although the depictions in Figure 1 include only in-plane lateral interstory drifts, they highlight the essential difference between conventional curtain wall systems and the proposed EICWS. Namely, vertical mullions in conventional curtain wall systems span more than one building story and are connected to the building frame at more than one story level, whereas vertical mullions in the proposed EICWS span only one building story and are attached to the building frame at only that particular story level. Thus, in conventional curtain wall systems, interstory movements in the building frame can cause curtain wall frame distortion and subsequent cladding panel damage (architectural glass panels, aluminum panels, concrete panels, etc.). In contrast, these same interstory movements should cause no damage in the proposed

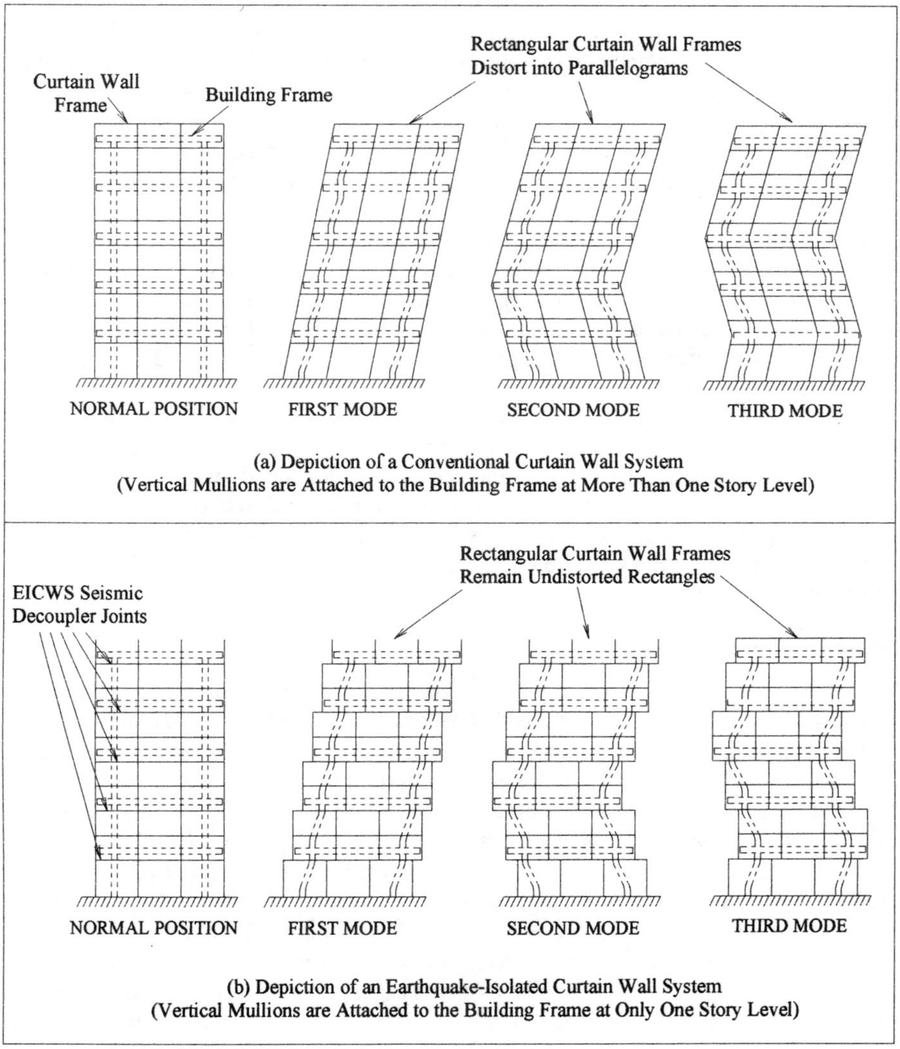

Figure 1 - *Schematic representation of fundamental vibration modes of a typical building frame: (a) clad with a conventional curtain wall system; and (b) clad with an Earthquake-Isolated Curtain Wall System.*

EICWS because of the "decoupling" that is achieved between adjacent stories in the EICWS frames.

Figure 2 depicts how a seismic decoupler joint [7] is able to accommodate relative interstory movements, while still maintaining a building envelope weatherseal. In-plane movements and out-of-plane movements are accommodated by horizontally continuous, flexible, elastomeric gasket loops, which act as weatherseals between stories. Thus, the seismic decoupler joint provides three, unimpeded translational degrees of freedom (X, Y and Z) between stories of a curtain wall system. The seismic decoupler joint also provides rotational decoupling between stories, but these degrees of freedom are less important in context of this application.

A face cap attached to the outside of the decoupler joint is free to rotate during out-of-plane movements due to its hinge connection and movement-accommodating sealant bead (Figures 2g and 2h). This face cap is installed primarily for aesthetic purposes, but it also acts as a water screen to prevent wind-driven rain from entering the building envelope. The face cap also provides protection from solar ultraviolet radiation for the elastomeric gasket loops.

One of the distinct advantages of a continuous seismic decoupler joint at each story level is the inherent resistance it provides to drift-related wall system damage, no matter which types of building plan or wall section shapes are employed (e.g., multi-story buildings with rectangular or irregular plan shapes, curved wall sections, set-backs, and re-entrant corners). The EICWS also gives the curtain wall designer more freedom to choose, for example, narrow mullion designs and less inherently drift-resistant glazing systems in earthquake-prone regions.

Laboratory Test Facilities for Evaluating the EICWS

In-plane dynamic racking crescendo tests on the conventional mid-rise curtain wall used as a basis for comparison with the EICWS were performed on the dynamic racking test facility shown in Figures 3 and 4. This test facility is described in greater detail by Behr and Belarbi [11]. Vertical mullions in the curtain wall test specimen were attached at all four corners to the facility's sliding steel tubes. These steel tubes slid on roller assemblies in opposite directions by means of a fulcrum and pivot arm mechanism. The bottom sliding steel tube was displaced by a computer controlled electrohydraulic servoactuator having a dynamic stroke capacity of ± 78 mm (± 3.1 in.). The fulcrum and pivot arm mechanism attached to the top and bottom sliding steel tubes doubled the effective servoactuator stroke capacity to ± 156 mm (± 6.1 in.).

The Dynamic Racking Test Facility shown in Figure 3 was modified to accommodate nearly full-height (due to laboratory ceiling height limits), one-story test specimens of the EICWS. Figures 5 and 6 show the dynamic racking test facility used for the EICWS tests. The nearly full-story EICWS specimens were secured to the middle sliding steel tube (representing the spandrel beam at "Story (i)" in the building frame) with specially designed steel subframes to support vertical mullions at only one story level in the primary building structure. The EICWS specimen decoupler joints were installed in the same manner as they would be installed on an actual multi-story building, with the exception that only one loop of the continuous flexible decoupler joint gasket was used because of a limited amount of gasket available for these tests. A more robust decoupler joint configuration would have two decoupler joint gaskets to provide a higher

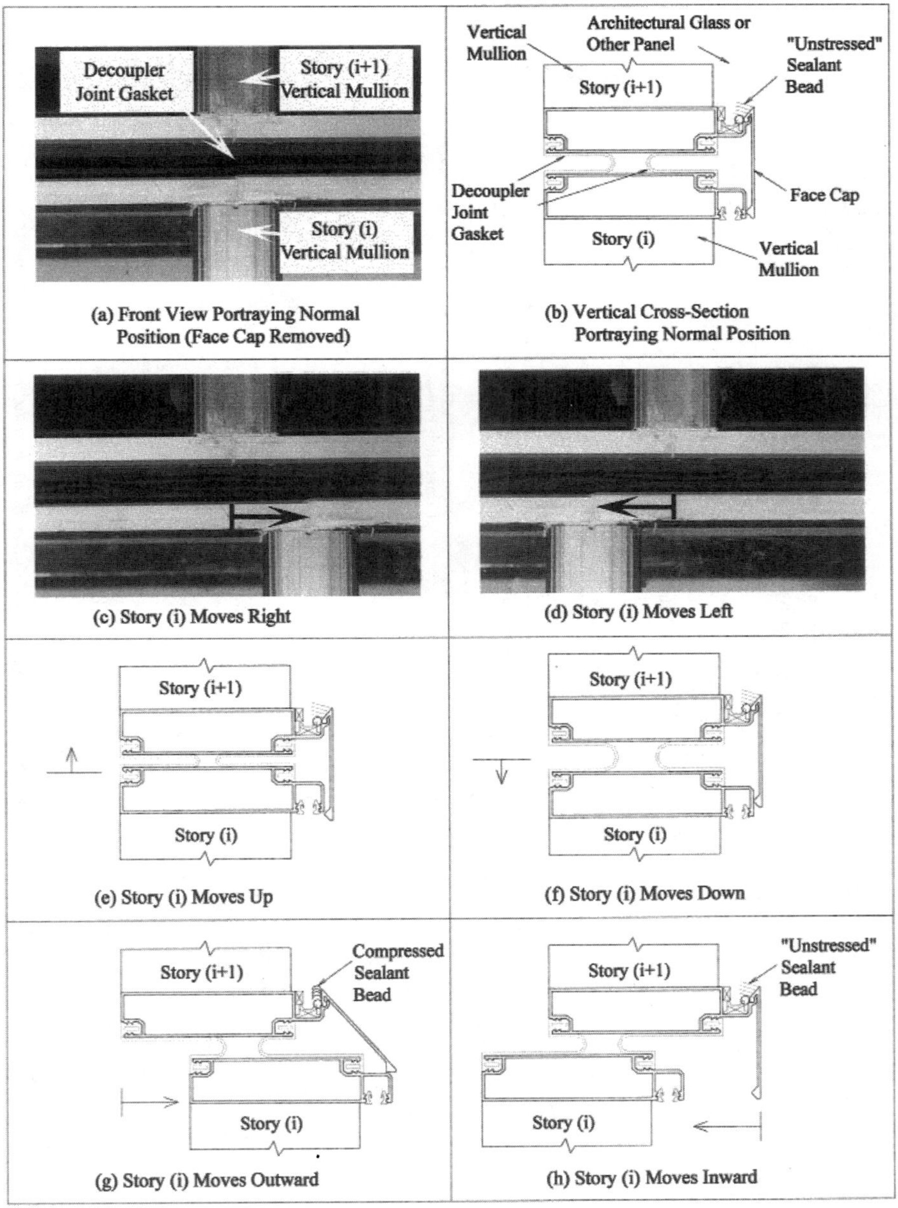

Figure 2 - *Accommodation of various building frame inter-story movements by the seismic decoupler joint of the Earthquake-Isolated Curtain Wall System: (a), (b) normal decoupler joint position; (c), (d) in-plane lateral movement about normal position; (e), (f) in-plane vertical movement about normal position; and (g), (h) out-of-plane movement about normal position.*

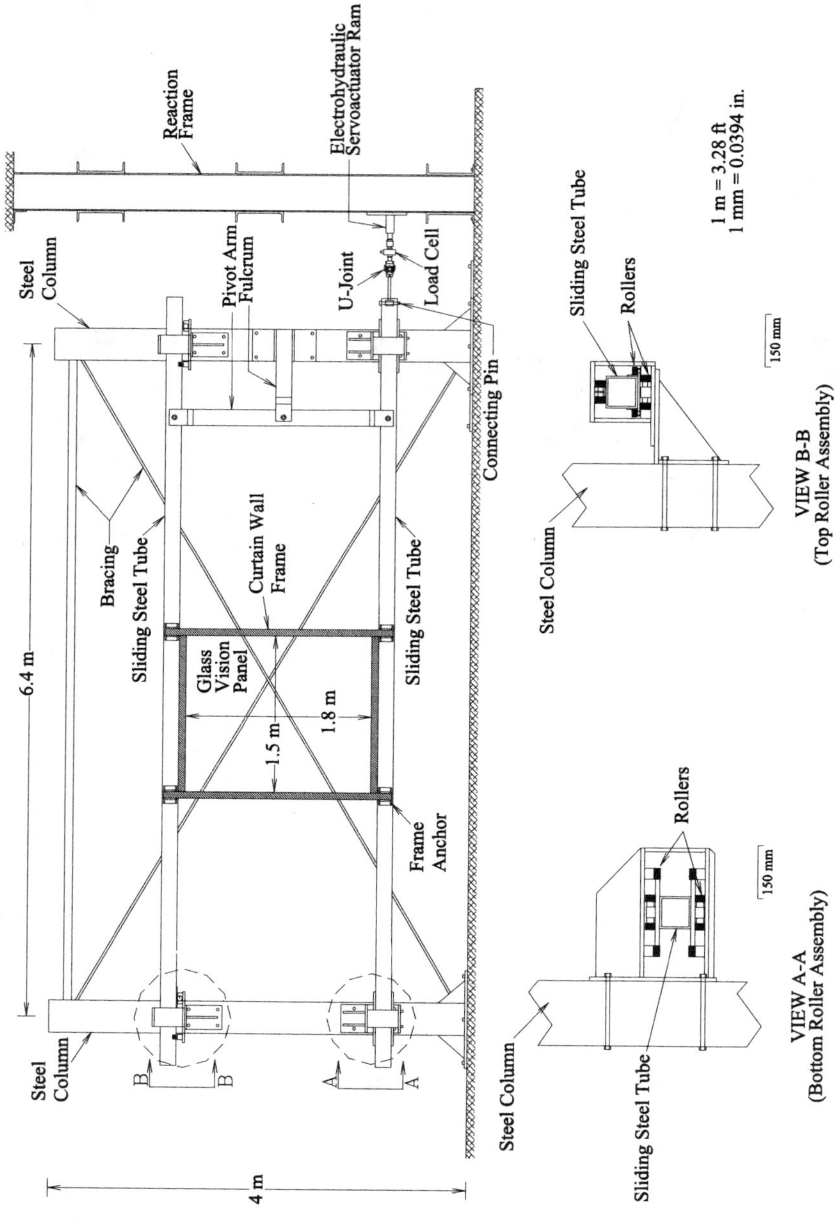

Figure 3 - *Facility used to perform dynamic racking tests on conventional curtain wall specimens.*

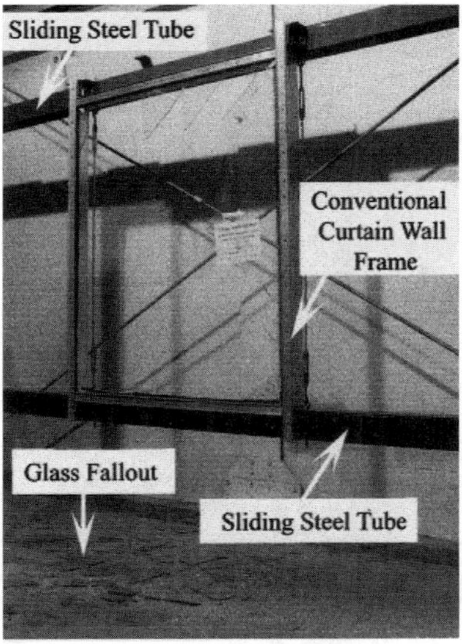

Figure 4 - *Photo of a conventional curtain wall system specimen attached to Dynamic Racking Test Facility.*

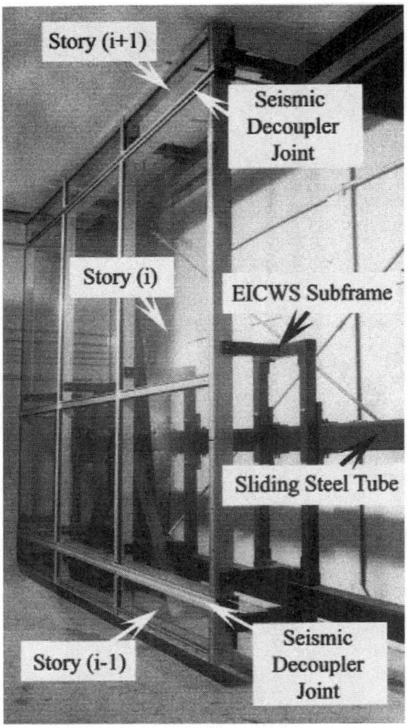

Figure 5 - *Photo of an Earthquake-Isolated Curtain Wall System specimen attached to Dynamic Racking Test Facility.*

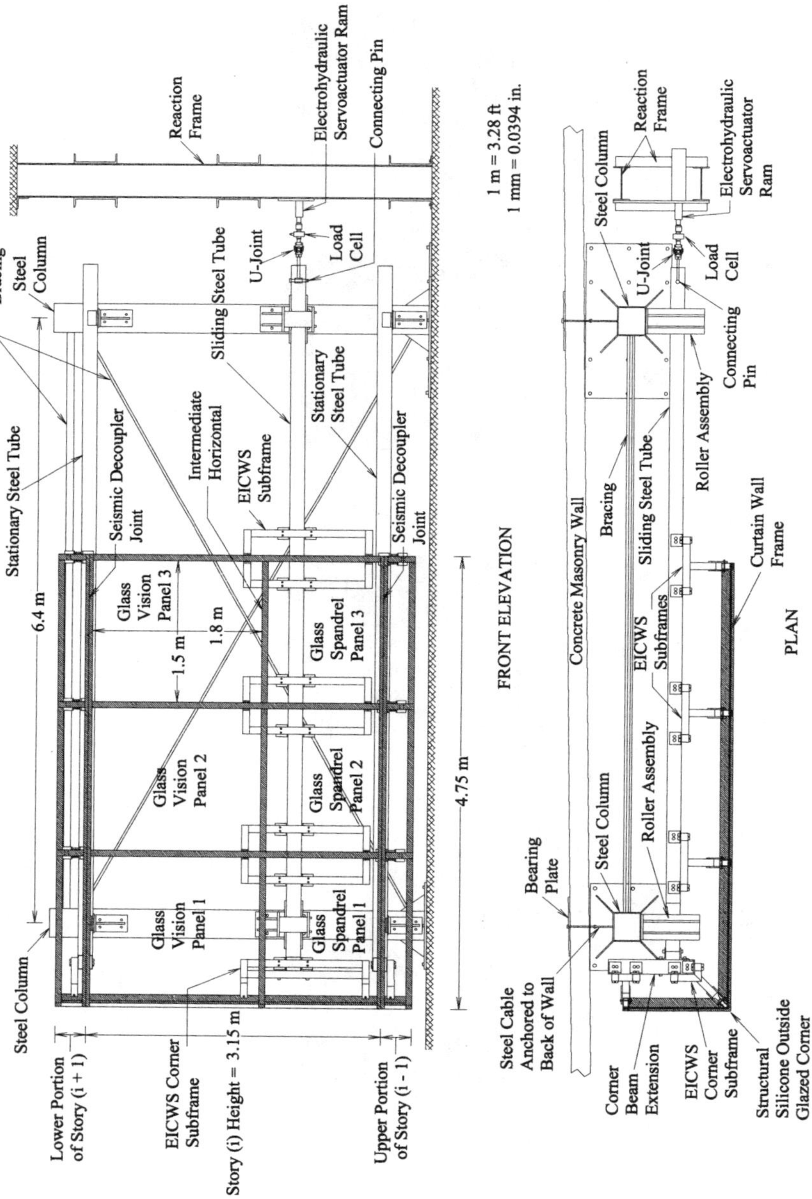

Figure 6 - *Facility used to perform dynamic racking tests on Earthquake-Isolated Curtain Wall System specimens.*

level of interstory building envelope weatherseal protection, as shown in Figure 2b.

The subframes used to anchor the EICWS specimens to the sliding steel tube were slightly different than those that would be used in an actual building installation. Specifically, each subframe's top anchor point was slightly above the curtain wall frame's intermediate horizontal for the EICWS specimens, whereas the top anchor point would be just below the intermediate horizontal in an actual installation to provide an unobstructed view from the building interior. In addition, it was necessary to construct the subframes with longer anchorage arms extending outward from the plane of the spandrel beams than would be used in practice. This was done in the EICWS specimens to ensure sufficient clearance between the dynamic racking test facility roller assemblies (Figure 5) and the curtain wall frame.

EICWS laboratory test specimens were comprised of an in-plane dry glazed section and a two-sided structural silicone glazed outside corner. The in-plane section was used for comparisons with the similarly glazed in-plane conventional curtain wall system specimens, whereas the outside glazed corner was used to explore the ability of the EICWS to accommodate out-of-plane interstory movements. Corner extensions (Figure 5) were used to attach the corner section of the EICWS specimen to the middle, sliding steel tube. Figure 7 details the curtain wall glazing elements used for both the conventional and EICWS specimens. Story (i) of the EICWS was "connected" to curtain wall story sections above and below using the seismic decoupler joints depicted in Figure 2. The adjacent story curtain wall sections [Story (i+1) and Story (i-1)] were attached to the upper and lower steel tubes of the dynamic racking test facility. Dynamic in-plane racking tests on the EICWS were performed by actuating the middle steel tube while the upper and lower steel tubes remained stationary; i.e., Story (i) moved relative to Story (i+1) and Story (i-1).

Air leakage rates through both the conventional and the EICWS specimen were measured using the portable air leakage test apparatus shown in Figure 8. ASTM Standard Test Method for Determining Rate of Air Leakage Through Exterior Windows, Curtain Walls and Doors Under Specified Pressure Differences Across the Specimen (E 283) was used as a procedural guideline for the measurement of air leakage rates through the curtain wall specimens. Test chambers were constructed by sealing 6 mil clear polyethylene shrouds to the specimens with duct tape. An inlet port for supplying pressurized air and an outlet port for measuring chamber pressure were attached to the plastic shroud. A pressure regulator connected to a compressed air source was used to supply air at the flow rate necessary to maintain the specified test pressure differential. Airflow into the test chamber was measured using one of three airflow meters over the 0 to 5.5 L/s (0 to 11.7 ft^3/min) capacity of the compressed air source, while the pressure differential across the specimen was monitored with a digital manometer. Chamber air conditions were monitored with a thermocouple and a barometer, whose readings were used to standardize the air leakage rates as prescribed in ASTM E 283-91.

Laboratory Test Procedures

The "crescendo test method" introduced by Behr and Belarbi [11] for evaluating the seismic performance of architectural glass and glazing systems was used as the basis for the dynamic in-plane racking tests performed on both the conventional and the EICWS

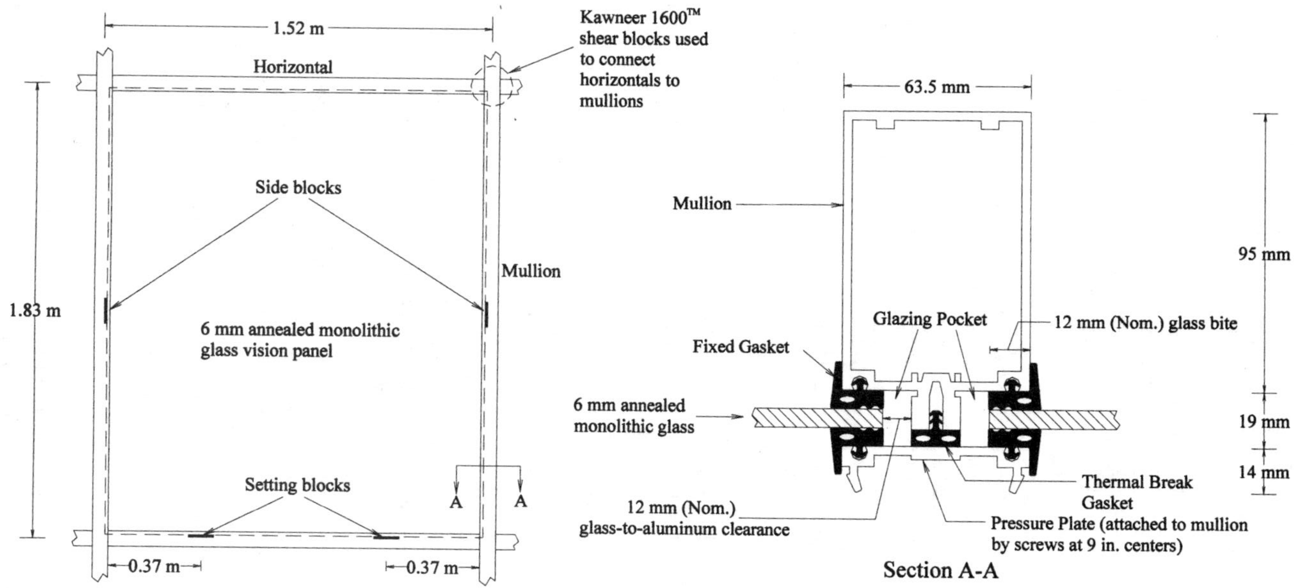

Kawneer 1600™ Glass-to-Frame Clearances:	
Top: 14 mm	Bottom: 10 mm
Left: 12 mm	Right: 12 mm

1 m = 3.28 ft
1 mm = 0.0394 in.

Figure 7 - *Dry-glazed Kawneer 1600™ framing system used for conventional and EICWS specimens.*

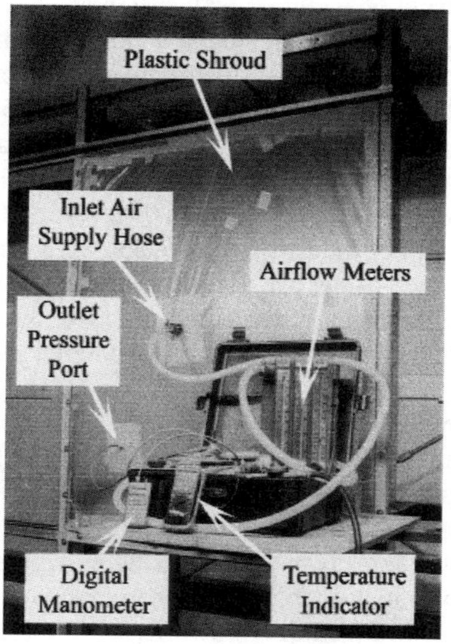

Figure 8 - *Photo of apparatus used to perform air leakage tests.*

specimens. The crescendo test method utilizes monotonically increasing sinusoidal drift amplitudes to determine serviceability drift limits and ultimate drift limits for architectural glass components subjected to dynamic, in-plane racking displacements. The crescendo test drift time history consists of a series of alternating "ramp up" and "constant amplitude" intervals, each comprised of four sinusoidal cycles at a nominal frequency of 0.8 Hz for hydraulic actuator strokes up to ± 38 mm (± 1.5 in.), and 0.4 Hz for actuator strokes greater than ± 38 mm. The crescendo test drift time history used in this study incorporated a pause in the dynamic racking after each constant amplitude interval in order to record pertinent wall system serviceability data (i.e., air leakage rate; glass, frame, or seal damage; etc.). A more detailed description of laboratory test procedures employed in this project is included in reference [12].

Visual inspections were performed during each crescendo test racking step pause to determine: (1) the "serviceability drift limit" defined by Behr [4] as the drift required to cause observable glass cracking (a condition that would necessitate glass replacement, but one that would not pose an immediate life safety hazard); and (2) the "ultimate drift limit" defined by Behr [4] as the drift required to cause glass fallout (a condition that could pose a life safety hazard to building occupants and pedestrians). A synchronized video recording of each dynamic racking test was used to provide a precise determination of serviceability drift limits and ultimate drift limits for the architectural glass test specimens.

Air leakage rates at selected drift indices were measured in accordance with ASTM E 283-91 for both the conventional and the EICWS specimens. However, the specimens did not correspond to the size prescribed by ASTM E 283-91, which specifies that the test section be at least a full building story in height and include both a vertical and a horizontal joint. Since the conventional curtain wall specimens in this study consisted of single vision panels of less than a full story in height, a modification in test section size was necessary. For the conventional curtain wall specimen air leakage tests, a single air leakage test chamber was used to envelope the single 1.5 x 1.8 m (5 x 6 ft) vision glass panel. During dynamic racking tests on the conventional curtain wall specimens, the plastic shroud used to create the air leakage test chamber was detached at the corners to allow for distortion of the curtain wall frame, and was then resealed prior to performing each subsequent air leakage test. For the EICWS air leakage tests, five air leakage test chambers were used: one over each of the two 1.5 x 1.8 m in-plane vision glass panels, one over the two corner vision glass panels combined, one over the seismic decoupler joint above the vision panels, and one over the seismic decoupler joint below the spandrel panels.

Isolating the in-plane glass panels, the corner glass panels, and both decoupler joints enabled investigation of air leakage rates in different zones of the EICWS as the dynamic drift amplitude increased. Plastic shrouds over the EICWS in-plane and corner vision panels remained sealed during dynamic racking displacements, since the EICWS curtain wall frame was not distorted during these in-plane racking movements. However, the plastic shrouds over the two seismic decoupler joints were removed for each crescendo racking step to prevent unanticipated restraint of the relative displacement between adjacent stories, which is accommodated entirely by the seismic decoupler joints (Figures 2c and 2d). Baseline air leakage tests for both the conventional and the EICWS specimens were performed before starting the dynamic racking tests, and were then repeated after predetermined crescendo racking steps (generally, every even numbered step) until the air leakage through the specimens exceeded the 5.5 L/s (11.7 ft^3/min) inlet airflow capacity of the air leakage test apparatus.

Laboratory Test Results

The same glass cracking serviceability drift limit of ± 38 mm (± 1.5 in.) [drift index of 1.9%] was observed during each of the three conventional curtain wall system crescendo tests. A slightly less repeatable (C.V. = 2.4%) average glass fallout ultimate drift limit of ± 64 mm (± 2.5 in.) [drift index of 3.1%] was observed. These drift limits are comparable to those reported by Behr [4] for dynamic racking crescendo tests performed on specimens of the same story height and also constructed with 6 mm (1/4 in.) annealed monolithic glass and Kawneer 1600™ curtain wall frames. Specifically, Behr reported an average glass cracking serviceability drift limit of ± 50 mm (± 2 in.) and an average glass fallout ultimate drift limit of ± 56 mm (± 2.2 in.). In contrast, no serviceability or ultimate limit states were reached in any of the three EICWS dynamic racking crescendo tests because glass cracking and glass fallout did not occur over the entire ± 152 mm (± 6 in.) [drift index of 4.9%] range of dynamic racking displacements.

Localized glass spalling and crushing were observed in the glass panel corners of the conventional curtain wall specimens as the panel corners were forced into contact with the aluminum frame members. Spalling and crushing were followed by cracks propagating radially outward from the corner regions of the annealed monolithic glass panels during subsequent crescendo racking steps, which soon led to glass fallout. No distortion of the curtain wall frame, nor movement of the glass panels within the frame, was observed during the EICWS dynamic racking tests, which explains the total absence of glass and frame damage during the EICWS tests.

Serviceability damage to the seismic decoupler joint gaskets of the EICWS was observed during only the latter stages of the dynamic racking crescendo tests. Seismic decoupler joint gasket damage was observed initially at drifts between ± 78 mm (± 3.1 in.) and ± 90 mm (± 3.5 in.) [± 82 mm average], which corresponded to drift indices from 2.5% to 2.8% (2.6% average). Decoupler joint gasket damage consisted of tearing at the corner splice of the flexible elastomeric gasket weatherseal installed between adjacent stories (Figure 2).

The prototype seismic decoupler joint used in this study was designed to accommodate a nominal out-of-plane movement of ± 51 mm (± 2 in.) before subjecting the gasket to tension. Test results revealed that this decoupler joint performed significantly better than its nominal design capacity (i.e., no damage occurred until ± 78 mm, well beyond the joint's ± 51 mm nominal out-of-plane movement capacity). In addition, the prototype decoupler joints in this study were able to accommodate drifts that were greater than the *NEHRP Provisions* (1997) prescribed allowable interstory drift for life safety of 0.02 times the story height (2% drift index) for structures that are not essential facilities or pose a substantial public hazard due to occupancy or use. For the 3.15 m (10.3 ft) story height used in the EICWS specimens, a 2% drift index would correspond to a ± 63 mm (± 2.5 in.) drift. Higher capacity seismic decoupler joint designs could employ a more flexible gasket material, an improved method to fabricate corner gasket splices, and a greater gasket loop length that would enable the EICWS to accommodate larger drifts without distress in the decoupler joint gasket.

Average air leakage rates, including ± 1 standard deviation error bars, are plotted against imposed drift index in Figure 9a for in-plane vision panel sections of both the conventional and the EICWS test specimens. A constant air leakage rate (i.e., no significant increase in air leakage rate) was measured for the three conventional curtain wall system vision glass panels up to a drift index of about 1.9%, at which time glass cracking was observed in each of the glass panels. Significant glass cracking occurred in the racking step immediately following the step associated with first observable glass cracking (i.e., at a drift index of 2.2%) in each of the conventional curtain wall test specimens. This cracking created large gaps for air to flow through during subsequent air leakage tests. Due to these gaps in the cracked glass panels, the air leakage test apparatus could only be used to make air leakage measurements on one of the three specimens tested at this racking step (the air supply used in the air leakage test apparatus could not maintain the specified pressure differential across the vision panel in the other two specimens due to the excessive air leakage through the panel). As shown in Figure 9a, no increase in air leakage was observed through the in-plane vision glass panels in the three EICWS test specimens over the entire ± 152 mm (± 6 in.) [drift index of 4.9%] range of dynamic racking, which greatly exceeds the life safety drift index limit of 2% prescribed in the *1997 NEHRP Provisions*.

Air leakage rates measured through the vision glass panels comprising the corner section of the EICWS specimens were similar to those measured through the in-plane vision panels in the same specimens. Corner section air leakage rates are plotted in Figure 9b and include the average air leakage rates through the two corner vision panels. Average air leakage rates for the seismic decoupler joints are plotted against drift index in

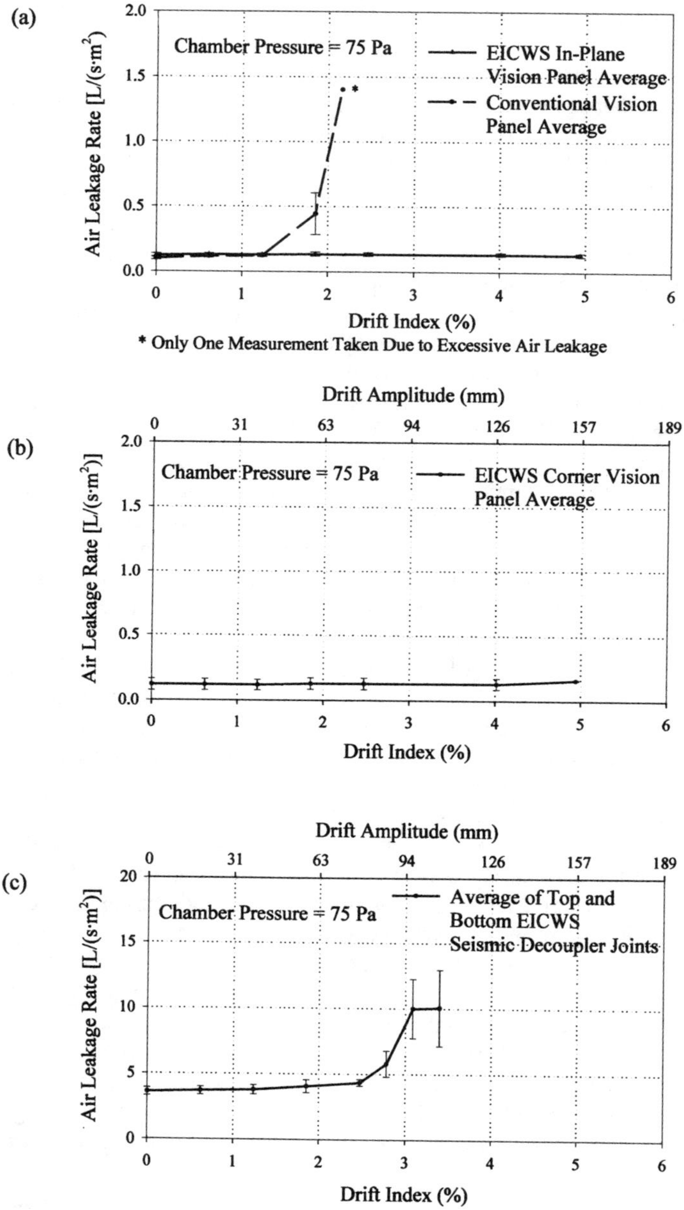

Figure 9 - Air leakage performance under in-plane dynamic racking drifts at a pressure of 75 Pa for: (a) conventional and EICWS in-plane vision panels; (b) EICWS corner vision panels; and (c) EICWS seismic decoupler joints. (1 mm = 0.0394 in., 75 Pa = 1.57 psf, 1 L/(s·m^2) = 0.197 cfm/ft^2)

Figure 9c. Again, it should be noted that the decoupler joint in this study was a prototype designed for a nominal out-of-plane movement: of only ± 51 mm (± 2 in.), which corresponds to a drift index of 1.6% for the EICWS test specimen. As shown in Figure 9c, air leakage rates did not increase significantly in the EICWS seismic decoupler joints until a drift index of 2.5% was reached, which is significantly higher than the *life safety* drift index limit of 2% prescribed in the *1997 NEHRP Provisions*. Thus, the EICWS test specimens exhibited no loss in serviceability — even when tested to levels that were representative of protecting life safety during severe earthquake events. A more detailed discussion of laboratory test results is included in reference [12].

Conclusions

First-generation seismic design provisions for architectural glass are slated for inclusion in the *2000 NEHRP Recommended Provisions for the Seismic Regulations for New Buildings and Other Structures* (to be published in 2001). These new seismic design provisions will increase the level of design attention paid to seismic life safety issues associated with architectural glass components in exterior wall systems.

In a separate endeavor, a new "Earthquake-Isolated Curtain Wall System" (EICWS) has been developed and tested successfully at the Penn State University Building Envelope Research Laboratory. By structurally decoupling each story of the wall system from adjacent story levels, the EICWS has shown an inherent ability to accommodate large interstory displacements in the vertical, horizontal and out-of-plane directions without jeopardizing life safety or compromising wall system serviceability. The EICWS has demonstrated an inherent ability to satisfy and exceed the seismic design provisions proposed for the *2000 NEHRP Provisions*. Not only was the EICWS completely immune to glass fallout, but it exhibited *serviceable* seismic performance at large imposed drifts that were certainly more representative of severe earthquakes.

Acknowledgment

Figures 1-9 and selected explanatory text related to these figures are reproduced from reference [12] with permission from the Earthquake Engineering Research Institute.

References

[1] Evans, D. and Ramirez, F. J. L., "Glass Damage in the September 1985 Mexico City Earthquake," *Lessons Learned from the 1985 Mexico City Earthquake*, Vitelmo Bertero (Ed.), Earthquake Engineering Research Institute, Oakland, CA, 1989, 233 pp.

[2] EERI, "Northridge Earthquake January 17, 1994: Preliminary Reconnaissance Report," ed. John F. Hall, Earthquake Engineering Research Institute, Oakland, CA, 1994, pp. 56-57.

[3] ICC, *International Building Code*, International Code Council, Falls Church, VA, 2000, Section 1621.1.

[4] Behr, R. A., "Seismic Performance of Architectural Glass in Mid-rise Curtain Wall," *Journal of Architectural Engineering*, Vol. 4, No. 3, 1998, pp. 94-98.

[5] Behr, R. A., Belarbi, A. and Brown, A. T., "Seismic Performance of Architectural Glass in a Storefront Wall System," *Earthquake Spectra*, Vol. 11, No. 3, 1995, pp. 367-391.

[6] BSSC, *1997 NEHRP Recommended Provisions for Seismic Regulations for New Buildings and Other Structures*, prepared by the Building Seismic Safety Council, Washington, D.C., for the Federal Emergency Management Agency and issued as *FEMA 302*, 1998.

[7] Wulfert, H. and Behr, R. A., "Earthquake-Immune Curtain Wall System," United States Patent Application Serial Number 09/093, 454, Filed June 8, 1998.

[8] American Architectural Manufacturers Association (AAMA), "Recommended Static Test Method for Evaluating Curtain Wall and Storefront Systems Subjected to Seismic and Wind Induced Interstory Drifts," Publication No. AAMA 501.4-2000, 2000.

[9] Bouwkamp, J. G. and Meehan, J. F., "Drift Limitations Imposed by Glass," *Proceedings of the Second World Conference on Earthquake Engineering*, Tokyo, Japan, 1960, pp. 1763-1778.

[10] Wright P. D., "The Development of a Procedure and Rig for Testing the Racking Resistance of Curtain Wall Glazing," Building Research Association of New Zealand (BRANZ) Study Report. No. 17, 1989, p.18.

[11] Behr, R. A. and Belarbi, A., "Seismic Test Methods for Architectural Glazing Systems, *Earthquake Spectra*, Vol. 12, No. 1, 1996, pp. 129-143.

[12] Brueggeman, J. L., Behr, R. A., Wulfert, H., Memari, A. M. and Kremer, P. A., "Dynamic Racking Performance of an Earthquake-Isolated Curtain Wall System," *Earthquake Spectra*, Vol. 16, No. 4, 2000, pp. 735-756.

Jeffrey L. Erdly and Robert W. Gensel[1]

A Basic Guide to Minimize Sealant Joint Failures in Exterior Building Walls

---

Reference: Erdly, J. L. and Gensel, R. W., **"A Basic Guide to Minimize Sealant Joint Failures in Exterior Building Walls,"** *Performance of Exterior Building Walls, ASTM STP 1422,* P. G. Johnson, Ed., ASTM International, West Conshohocken, PA, 2003.

Abstract: While the quality and performance of elastomeric sealants used in exterior building walls have improved, the occurrence of leakage and envelope failures attributed to sealant installation has not decreased. This paper will outline, from a contractor's viewpoint, what designers need to know with regard to sealant joint design and detailing to minimize building wall leakage attributed to sealant joint failures.

Keywords: sealants, joint design, interface, adhesion

As building cladding systems have evolved, their reliance on the effective installation of elastomeric sealants to maintain and/or minimize water infiltration has increased. Mass masonry walls, which were prevalent in construction through the early 1950s, did not rely on sealants as a primary buffer to stop water infiltration. The redundant design of these massive wall assemblies allowed for the accommodation of some moisture within the system. Subsequently, with the advent of the curtain wall, high performance elastomeric sealants were required to seal these modern cladding systems where joints experienced rapid cyclical movement and sealants were intended to provide a primary weather seal between facade components. The design of these wall assemblies relies on the sealant joints to inhibit water infiltration to the building. Sealant materials and, theoretically, joint design had, therefore, to evolve to meet the needs of these new wall systems.

While the materials used to seal construction joints have evolved, many designers still overlook the need to design a functional sealant joint which is, not only, constructible, but which will also provide long term durability. Additionally, the sealant contractor is generally the last member of the construction team on the project. Even if the joints have been properly designed, construction tolerances or material substitutions may significantly impact the as-built joint width, resulting in an ill-advised joint profile. Near the end of a construction project, the last thing the general contractor, architect or owner wishes to hear is a sealant contractor rejecting the as-built joint configurations as inadequate. Zwayer et al [*1*] identifies many common errors that occur at sealant joint interfaces.

---

[1] President and Sr. Vice President respectively, Masonry Preservation Services, Inc., P.O. Box 324, Berwick, PA 18603.

The New Construction Sealant Industry

Because of the difficulties associated with new construction sealants, many qualified specialty contractors have withdrawn from the new construction market, and work only as remedial repair contractors. Mr. George Grenadier, acknowledged as the pioneer in the use of exotic (early polysulfide), high performance sealants in the United States, made the following statement in 1976, [*2*] "For a number of years we have not participated in new construction projects. Instead, all of our operations are devoted to remedial work - problem solving." Many of the problems identified by Mr. Grenadier beginning in 1952 still plague the new construction sealant industry today.

On a project located in northeastern Pennsylvania, a specialty contractor rejected many of the as-built joint configurations as they were undersized and the aluminum window frame extrusions had been fabricated with no return leg. This effectively resulted in the lack of a suitable substrate for the installation of a backing material or for the sealant to adhere to. Per the specification, the specialty contractor proceeded to forward a letter to the general contractor rejecting the as-built joints as noncompliant, based on a caveat contained in the specification, stating that by commencing the installation of sealant, the contractor accepted the as-built joints as acceptable. After a heated job meeting with both the general contractor and the architect, the GC was directed to "find a sealant subcontractor who did not write letters." Two years later, the specialty contractor was contracted to install remedial sealants to stop the water infiltration at this project, at a cost substantially higher than his original bid.

Another factor, which contributes to the failure of the sealant joints, is competitive pricing, whereby highly skilled, competent contractors cannot compete in the market. When a specialty contractor was asked by an established client to price a new construction project to their general contractor, they were told the following: "We will consider your number, but generally we get the window sub to install his sealants, the storefront people handle their work and when we contract with the masonry subcontractor, they generally 'throw in' their sealants as part of the negotiation." The statement 'throw in' implies that little or no cost is included, however, the problems associated with failed sealant joints and a leaky building quickly negate any potential cost savings.

The State of the Art

The ASTM Specification for Elastomeric Joint Sealants C920-98 and Guide for Use of Joint Sealants C1193-91 (re-approved 1995) [*3*] provide a complete, succinct guide for the selection of sealants and the proper design of joints. Additionally, O'Connor [*4*] Myers [*5*] and Nicastro [*6*] provide in-depth design guidance for many atypical sealant joint configurations. Also available to the designer of sealant joints is a substantial volume of technical papers regarding wall leakage caused by sealant joints. Samplings of these include Kudder and Lies [*7*] and Warseck [*8*].

Many new buildings are clad with exterior insulation and finish systems (EIFS). Since most EIFS clad structures built to-date intended the cladding as a barrier wall whose success is largely dependent on sealant success, functioning long lasting sealant joints are more important than ever. See Williams, M.F. and Williams, B.L. [*9*] *Exterior Insulation and finish systems: Current Practices and Future Considerations.* For additional information regarding EIFS and water penetration, see Kenney and Piper 1992, Williams and Williams 1990, Nelson and Waltz 1996, and Chin et al. 1999 [*10-13*].

Joint Design

Generally, the successful installation and performance of any sealant project begins with a proper joint design. As noted, numerous technical bulletins, along with published ASTM guidelines, are readily available to the design professional and are invaluable with regards to joint configurations, sizing and the selection of the appropriate sealant. With today's wide array of available cladding materials, along with the continued practice of identifying less costly veneer alternatives, proper joint design becomes increasingly more imperative. These wall assemblies not only present unique challenges with regards to sealant adhesion and joint serviceability, but also, are typically not properly designed to dissipate water that may breach the surface.

It is advisable during the design phase to develop full-scale details of each joint configuration which will be encountered on the project. Problems or concerns which may not be evident within a typical ½″ or ¾″ scale detail often become glaringly obvious when detailed at the size at which they will be constructed. Insufficient (or lack of) return legs on aluminum frame components, for example, are quite evident when viewed at actual size. Identification of such concerns, such as adhering sealant to a _″ frame edge during design, would allow for necessary revision of either the frame component or the adjacent wall assembly to permit a proper sealant joint configuration. Furthermore, by identifying and addressing insufficient or problematic joint configuration prior to construction, "field revision" of sealant details by the contractor will be of less concern.

Upon review of the proposed joint details and following the necessary revisions to ensure a proper configuration, hands-on testing of each joint type should be conducted. Generally, full size cross-sectional mock-ups of the various components to be utilized in construction should be considered. These mock-ups should include not only the actual dimensions of the components to ensure the viability of the installation, but also the actual materials to be utilized to confirm acceptable adhesion of the proposed sealants. Even the most carefully designed joint will ultimately fail if the sealant selected for installation does not adhere to or is not compatible with one or more of the substrates encountered within the project. For example, although many of today's silicone sealants achieve acceptable, unprimed adhesion to many substrates, adhesion to some of today's flouropolymer coatings used on aluminum frame components is problematic for some of these materials. In this case, the use of primers or potentially a change in sealant would require consideration.

Invariably, pre-construction testing of both joint configurations and materials identifies one or more deficiencies which may be detrimental to sealant joint performance on a particular project. It is far less complicated and always less costly to modify the design or proposed material prior to construction than it will be after the contractor is mobilized onsite and a delay is encountered while options are explored. Most always, the resulting change order reflects not only the designer's costs to address the modification, but also the contractor's costs (real or perceived) related to production delays and/or scope changes. It is interesting to note that when this idea is discussed with many design professionals, the typical response is that there is not enough time or resources available to them to conduct the proposed testing. Perhaps the cost of preliminary testing should be weighed against the cost of potential moisture-induced deterioration, which may result when sealant joints fail prematurely.

Contractor Qualifications

The overall performance of any sealant joint is, generally, in direct correlation to its original design. However, proper design alone does not ensure a successful sealant project. The installation of the sealant material plays as important a role as the design itself. Far too often a contractor, either inexperienced or who disregards proper installation techniques, is retained on a project based solely on a "low bid." Many times, the resulting sealant installation may prove disappointing and costly, both during construction and throughout the service life of the building.

It is often difficult to identify a qualified sealant contractor on a particular project. It seems that, more and more, any contractor involved in exterior facade work, including window washers, and cleaning contractors believe themselves to be qualified to perform sealant installation. Although some of these firms may indeed be qualified, by and large, the number of inexperienced firms dwarfs the number of highly qualified firms. Furthermore, as many projects rely basically on the "low-bid" approach to retain a contractor, the vast majority of new construction sealant installations are awarded to firms who may not be capable of providing the service that the owner and professional believe they are receiving. Often times, the remedial repair of these improperly installed joints becomes necessary much sooner than anticipated. Unfortunately, at this point, additional repair scopes to address corroded fasteners, and or interior damage, which have resulted from water infiltration are also necessary. These costs, along with the remedial repair itself and, in some cases, litigation against the original installer, make the savings realized by retaining the "low bid" installer seem rather insignificant.

It is highly improbable that on every project, a competent installer will be identified and retained. There will continue to be a faction within the building industry who bases contractor selection on the lowest cost and, on these projects, the probability of retaining a qualified installer is based on "the luck of the draw." However, for owners, construction managers, and designers who recognize the value of investing in a qualified contractor in order to minimize repairs and maximize service life, there are options available to them. A sampling of a few techniques and methods, which may, in our experience, help to identify a qualified installer follows:

- The designer should, prior to considering costs, pre-qualify the sealant contractors who will be asked to quote the project. In addition to contacting former clients, owners and designers regarding their satisfaction with the installer, the competing contractors may also be of help in this regard. Most qualified installers have no objection to quoting projects against other similarly qualified firms. They will, however, be uneasy about competing against a less competent, "low bid" firm.

- Conduct a face-to-face interview with the attractive bidders on a project. Many times, the expertise (or lack thereof) of a contractor will be evident upon discussions regarding the project. Specific questions regarding any difficult, or atypical joint types that may be encountered should be asked.

- Discuss the technical aspects of the project with a project manager, foreman, technician or other such representative who will be directly involved on-site. It has been our experience that, those firms that direct all inquiries to a "salesman" may be compensating for a lack of technical knowledge below the management level. Keep in mind that the "salesman" will not be performing the hands-on installation on the project. Unfortunately, in many of these firms the field employees who install the sealants may lack the experience or understanding necessary to accomplish a high quality application.

- Inform the contractors during the bid stage that random, onsite testing of installed sealants will be conducted throughout the project. Direct the contractor to include the cost to repair these test locations in their quote. It is imperative, however, that this field testing then be conducted. Many designers use the threat of field tests to keep the contractor honest, however, the reputation of these firms who do not actually follow through on these inspections, quickly travels throughout the industry.

- Upon selection of a contractor and following commencement of the project, visit the site. Inspect some of the technician's tool buckets to confirm that they possess the right tool for the job. It may seem odd that the technicians possess only a ½″ slicking tool on a project that entails a 1″ wide precast joints. This obviously should be a red flag.

- Most highly skilled, responsible contractors also perform onsite adhesion testing to ensure that an acceptable bond between the sealant and the substrate continues to be achieved. Although original testing may confirm acceptable adhesion, circumstances may occur which negatively affect this bond. It is not uncommon that circumstances such as modifications to coatings on frame components, residual form release or road grime on a shipment of precast units, or other such anomalies may occur which could impact adhesion. Although preconstruction testing may be successful, substrate changes which may lead to inadequate adhesion of the new sealant may not be identified. In extreme instances, widespread, premature failure of these sealant joints could be anticipated.

- Make every effort to personally meet and observe each mechanic on the project who will be involved in the installation of the new sealants. Although the foreman and one or two mechanics may appear competent, they will not be installing sealant personally at each location. Although skill levels will always vary between each mechanic, it is important that, at a minimum, each installer has a working knowledge of the proper techniques required for a successful installation.

The installed cost of elastomeric sealants on any project often pales in comparison with the other components contained in a wall assembly. On a recent window replacement project, 5% of the project cost was attributable to the sealant installation. Additionally, as the design of these assemblies has evolved to the typical cavity or curtain wall design currently used, they have become more sensitive to the infiltration of significant water due to leakage. The installation of sealants which prematurely fail, whether by inadequate design or improper installation, will ultimately lead to costly, labor-intensive repairs, including replacement of ferrous metal components, saturated insulation and sheathing, which possibly could have been prevented. The entire project team, beginning with the owners and including the designer, construction managers and contractors, must grasp the concept of the methodical, state-of-the-art requirements of today's sealant installations. Perhaps the next time a contractor offers to "throw in" the sealants as part of a package, the long-term ramifications, as well as the true cost of this low bid item, may be more apparent.

References

[1] Zwayer, G. L., Cook, D. F., and Crowe, J. P., "Leakage at Interface Between Multiple Facade Components," *Water Leakage Through Building Facades, ASTM STP 1314,* R. J. Kudder and J. L. Erdly, Eds., ASTM International, West Conshohocken, PA, 1998, pp. 142–154.

[2] Grenadier, George, "A Sealant Applicator's Viewpoint," *Building Seals and Sealants, ASTM STP 606,* ASTM International, West Conshohocken, PA, 1976, pp. 151–161.

[3] Annual Book of ASTM Standards, Vol 04.07, ASTM Standard Specification for Elastomeric Joint Sealants C920-98, and Standard Guide for Use of Joint Sealants C1193-91 (re-approved 1995).

[4] O'Connor, T. F., "Design of Sealant Joints," *Building Sealants: Materials, Properties and Performance, ASTM STP 1069,* Thomas F. O'Connor, Ed., PA, 1990.

[5] Myers, J. C., "Behavior of Fillet Sealant Joints," *Building Sealants: Material Properties and Performance, ASTM STP 1069,* Thomas F. O'Connor, Ed., ASTM International, West Conshohocken, PA, 1990.

[6] Nicastro, D. H., "Difficult Sealant Joints," *Science and Technology of Building Seals, Sealants, Glazing, and Waterproofing: Second Volume, ASTM STP 1200,* Jerome M. Klosowski, Ed., ASTM International, West Conshohocken, PA, 1992.

[7] Kudder, R. J. and Lies, K. M., "Diagnosing Window and Curtain Wall Leaks," *Water in Exterior Building Walls: Problems and Solutions, ASTM STP 1107,* Thomas A. Schwartz, Ed., ASTM International, West Conshohocken, PA, 1991.

[8] Warseck, K. L., "Why Construction Sealants Fail - An Overview," *Building*

Sealants: Materials, Properties, and Performance, ASTM STP 1069, Thomas F. O'Connor, Ed., ASTM International, West Conshohocken, PA, 1990.

[9] Williams, M. F. and Williams, B. L., *Exterior Insulation and Finish Systems: Current Practices and Future Considerations,* ASTM Manual Series: MNL 16, ASTM International, West Conshohocken, PA, 1994.

[10] Kenney, R. J. and Piper, R. S., "Factors Affecting the Durability of Sealants in Contact with the Finish Coating of EIF Systems," *Science and Technology of Building Seals, Sealants, Glazing, and Waterproofing, ASTM STP 1168,* C. J. Parise, Ed., ASTM International, West Conshohocken, PA, 1992, pp. 117–127.

[11] Williams, M. F. and Williams, B. L., "Sealant Usage for Exterior Insulation & Finish Systems," *Building Sealants: Materials, Properties, and Performance, ASTM STP 1069,* Thomas F. O'Connor, Ed., ASTM International, West Conshohocken, PA, 1990.

[12] Nelson, P. E. and Waltz, M. E., Jr., "EIFS - Surface-Sealed Wall Systems That Need Flashings," *Exterior Insulation Finish Systems (EIFS); Materials, Properties, and Performance, ASTM STP 1269,* Peter E. Nelson and Richard E. Kroll, Eds., ASTM International, West Conshohocken, PA, 1996.

[13] Chin, I. R., Thompson, T. S., and Rouse, B. K., "Exterior Insulation and Finish Systems (EIFS) Water Resistance/Leakage," *Water Problems in Building Exterior Walls: Evaluation, Prevention, and Repair, ASTM STP 1352,* J. M. Boyd and M. J. Scheffler, Eds., ASTM International, West Conshohocken, PA, 1999.

Konopka, K. M.,[1] *McKelvey, J. L.,*[2] *Rimmer, J. W.,*[3] *O'Brien, M. J.*[4]

Selected Performance Characteristics of a Dual Purpose 100% Acrylic Polymer-Based Coating that Performs as Both a Weather Resistive Component for Exterior Insulation Finish Systems (EIFS), and as an Adhesive for Attachment of the Insulation Board

---

Reference: Konopka, K. M., McKelvey, J. L., Rimmer, J. W., O'Brien, M. J., **"Selected Performance Characteristics of a Dual Purpose 100% Acrylic Polymer Based Coating that Performs as Both a Weather Resistive Component for Exterior Insulation Finish Systems (EIFS), and as an Adhesive for Attachment of the Insulation Board,"** *Performance of Exterior Building Walls, ASTM STP 1422*, P. G. Johnson, Ed., ASTM International, West Conshohocken, PA, 2003.

Abstract: The multi-component Exterior Insulation Finish Systems (EIFS) designed nearly 30 years ago are still the basis for today's EIFS. However, over the years new EIFS have been developed to meet the needs of the building industry. One such system, EIFS with drainage, incorporates the components of the basic barrier EIFS, but adds the dimension of a drainage plane behind the insulation board. One EIFS with drainage system consists of a liquid coating that can be used as a flexible weather resistive barrier in place of building paper, and as an adhesive in place of the metal fasteners

This paper describes the testing and selected performance characteristics of one component of EIFS with drainage, an acrylic polymer-based coating that trowels easily, forms breathable, highly water resistant weather barrier assemblies, and adheres the insulation board to the typical EIFS sheathings, all in one step. Early grab, or "green strength", helps hold the insulation board when it is first set in place, which may allow the insulation board to be rasped sooner than when a more conventional liquid adhesive is used.

Full component testing of the EIFS, including joint-treatment and flashing requirements, is not the objective of this paper.

Keywords: acrylic, durability, EIFS with drainage, Exterior Insulation Finish Systems (EIFS), barrier EIFS, weather resistive barrier, weather resistive coating

---

[1]Senior Research Scientist, [2]Research Scientist, [3]Group Leader, [4]Section Manager, Coatings Technical Service, Rohm and Haas Company, 727 Norristown Road, Spring House, PA 19477.

Introduction

Exterior Insulation Finish Systems (EIFS) have been used successfully for more than thirty years in the United States and Canada. At the present time, EIFS have approximately 20% of the total commercial cladding market, and residential EIFS are growing due to their attractive curb appeal and insulation value to the consumer. The multi-component system designed nearly 30 years ago is still the basis for today's EIFS. However, over the years new innovative EIFS have been developed to meet the needs of the building industry. One such system is EIFS with drainage.

EIFS with drainage incorporates the components of the basic barrier EIFS, but adds the dimension of a drainage plane behind the insulation board. The drainage plane allows incidental water that may get to the backside of the system to be directed to the outside of the wall assembly by the means of a weep screed at the bottom of the wall. In one such drainage system, building paper and mechanical fasteners have been used to protect the sheathing and fasten the insulation board to the substrate, respectively. In more recent years, alternative systems have been developed to replace the building paper and fasteners. One such system consists of a liquid dual-purpose coating that can be used as a flexible weather resistive coating (WRC) in place of building paper, and as an adhesive in place of the metal fasteners.

The International Conference of Building Codes (ICBO) was founded in 1922 for the development of a building code that all communities would accept and enforce. It currently publishes the *Uniform Building Code*. Trowel applied weather-resistive coatings are included in the ICBO, AC 24 *Acceptance Criteria for Exterior Insulation and Finish Systems* (Effective November 1, 1999).

Acrylic polymers have been used for nearly 25 years in EIFS as modifiers and binders to formulate adhesives, basecoats, and finishes. Recently, an acrylic polymer was introduced commercially for use in formulating emulsion based, trowel applied, liquid, weather resistive coatings (WRC)/adhesives. This soft binder (glass transition temperature, or Tg, of –35 C) provides the durability and flexibility of an acrylic, and it also introduces rheology that provides coatings that trowel easily, form breathable highly water resistant films, and adhere the insulation board to the typical EIFS sheathings, all in one step. Early-grab helps hold the insulation board when it is first set in place, prior to rasping. Rasping of the insulation board is followed by application of the base coat and reinforcing mesh, and then a finish coat.

This paper describes testing and performance of this WRC according to all tests of ICBO AC 24, with the exception of water penetration as measured by *ASTM Test Method for Structural Performance of Exterior Windows, Curtain Walls, and Doors by Cyclic Static Air Differential* (E 1233), and *ASTM Test Method for Water Penetration of Exterior Windows, Curtain Walls, and Doors by Static Air Pressure Difference* (E 331). This water penetration testing has not been performed by the authors because the requirements of E 1233 and E 331 exceed the physical capacity of our laboratory. It is anticipated by the authors that the WRC would pass this test, however this has not yet been demonstrated.

Test results show the WRC/adhesive based on the acrylic binder to pass the selected tests listed for EIFS weather resistive coatings, as outlined in ICBO AC 24. These acceptance criteria measured in our laboratory are outlined below.

ICBO AC-24 Section	Test	Required
6.4.1	Tensile Bond: plywood 'mesh faced' gypsum board 'paper faced' gypsum board	Minimum of > 15 psi.
6.5.1	Freeze-thaw: plywood 'mesh faced' gypsum board 'paper faced' gypsum board	no surface changes that effect performance.
6.6.1	Water resistance: plywood 'mesh faced' gypsum board 'paper faced' gypsum board	absence of deleterious effects.
6.8.1	Water vapor transmission of materials: perms gms/24 hrs/m^2	UBC Standard 14-1, Table 14-1-A.

In addition to the testing required by the ICBO AC 24 Acceptance Criteria, there are several other laboratory tests that help us to characterize a WRC. These additional tests, listed below, will be discussed in more detail in the body of this paper.

- Percent Elongation
- Resistance to Hydrostatic Pressure
- Accelerated Weathering
- Early Grab.

Coating raw material quality, overall coating composition, and manufacturing techniques can effect the performance of a formulated WRC. Therefore, a commercial WRC would require full testing in order to assure acceptable performance under the ICBO AC 24 Acceptance Criteria.

Polymer type

A description of acrylic polymers has been reported previously in some detail [*1,2*]. The properties of acrylic polymers, including Tg, that make them desirable binders for use in EIFS have also previously been reported [*3*]. This acrylic binder was specifically developed for use in dual purpose WRC/adhesives designed for EIFS with drainage. It provides rheology which allows the WRC/adhesive to be easily troweled, and to develop early grab, sometimes referred to as "green strength".

Formulations and application

The formulation contains acrylic emulsion, defoamer, pigment dispersant, surfactant, water, in-can preservative, thickener, extenders, and sand aggregate. The ingredients are mixed on a laboratory dough mixer for about ten minutes.
The sand used in the formulation acts as a thickness guide during application and provides sufficient film build to attain the suggested dry film thickness. The formulation described in this paper incorporated 60 mesh sand which provided a film thickness of approximately 18 mils. Commercial formulations may vary in both component types and levels resulting in different applied thickness and film-thickness requirements.

The WRC/adhesive is trowel applied to the sheathing and the Expanded PolyStyrene board (EPS) is firmly placed onto the wet WRC/adhesive. The EPS board is rasped once the adhesive has cured sufficiently to prevent board movement. Rasping is followed by application of EIFS basecoat and finish.

The formulation parameters used to evaluate the WRC/adhesive are listed below.

<u>WRC Formulation Parameters</u>

Pigment Volume Concentration	49.0%
Volume Solids	64.9%
Weight Solids	77.6%
Viscosity, Paste Units	120 - 135
pH	7.0 - 7.6

Test procedures

ICBO AC 24 Acceptance Criteria

Application advantages must be combined with the ability to meet ICBO AC24 performance standards in any new product designed for use in EIFS applications. This dual-purpose WRC/adhesive has a smooth "buttery" feel under the trowel designed to minimize applicator fatigue. The coating also meets the selected sections of the ICBO, AC 24 Acceptance Criteria which were laboratory tested. A brief description of the ICBO AC 24 tests included in our study, along with test results for the dual purpose WRC/adhesive follows.

ICBO AC 24, Section 6.4.1: Tensile Bond - Tensile Bond testing measures the strength of an EIFS composite under tension. The ICBO AC 24 Acceptance Criteria requires a minimum flatwise tensile strength of 15 psi. for an EIFS composite incorporating a water resistive coating. Testing must comply with the *ASTM Test Method for Tension Test of Flat Sandwich Construction in Flatwise Plane* (C 297). Table 1 shows that a composite system of substrate/WRC/EPS provides at least 15 psi when using plywood and mesh-faced gypsum board substrates. These composites were prepared by trowel applying the WRC/adhesive to the substrate and placing the EPS directly onto the wet WRC/adhesive. The composites were dried for both 1 and 7 days at 25°C/50% relative humidity (RH) prior to testing. All failures during Tensile Bond testing were in the EPS board, indicating that the tensile strength of the WRC-to-substrate bond, and WRC-to-EPS bond is at least 15 psi. The failure mode in these tests was the cohesive strength of the EPS board.

Table 1 - *AC-24 Tensile Bond Adhesion, ASTM C 297 (substrate/WRC/EPS composites)*

Substrate	**Tensile Strength, psi.**	
	24 hr cure	1 wk. cure
Plywood	15	15
Mesh-faced gypsum board	17	15

Note: All failures were Cohesive in the EPS.

In a second Tensile Bond test, we prepared composites by trowel applying the WRC/adhesive to mesh-faced gypsum board, paper-faced exterior gypsum board, and plywood. The samples were cured for 7 days at 25°C/50% RH, and then a wooden block was glued to the cured WRC to facilitate Tensile Bond testing. Table 2 shows the results of this test. Tensile strength was 28 psi. for the WRC-to-mesh-faced gypsum board composite, and 25 psi for the WRC-to-paper-faced exterior gypsum board composite, with substrate failure in both cases. Tensile strength for the WRC-to-plywood composite was 63 psi, and the failure mode was cohesive in the WRC. These results show that the mesh faced gypsum board and paper-faced exterior gypsum board substrates have higher cohesive strength than the EPS board (see Table 1), but they have significantly lower cohesive strength than that of the subject WRC.

Table 2 - *Tensile Bond Adhesion, ASTM C 297 (WRC/substrate composites)*

Substrate	**Tensile Strength, psi.**
Plywood	63, C in WRC
Mesh-faced gypsum board	28, C in substrate
'Paper-faced exterior gypsum board	25, C in substrate

C=Cohesive failure.

ICBO AC 24, Section 6.5.1: Freeze-thaw - Freeze-thaw (F/T) testing (Table 3) evaluates the performance of EIFS incorporating a WRC under conditions of total water immersion followed by freezing temperatures. A F/T cycle consists of a minimum of 8 hours at 49° C , followed by 8 hours of total immersion in water at 21.1 °C to 26.8 °C, followed by 16 hours of freezing at minus 28.9 °C Ten cycles are required by the ICBO AC 24 Acceptance Criteria.

Test panels (five required) must be 23226 mm^2 (6-inch squares) that are cut in half and then coated with the WRC. The sides and back of the panels are sealed with a water impervious material so water can only enter the panel through the WRC. In our testing, we trowel applied the WRC to the substrate, cured the WRC for two days at room temperature, and then used epoxy glue to seal the seam on the back and edge of the test panel. The back and edges of the panels were also sealed using two brush-applied coats of a commercial alkyd paint.

In the ICBO AC 24 Acceptance Criteria, failure is defined as surface changes such as cracking, checking, crazing, erosion, or other characteristic that may affect performance as a wall cladding. Failure is also defined as delamination, or indications of delamination between components, as viewed under a minimum of 5X magnification.

The WRC/Adhesive based on the acrylic binder passes this F/T testing, showing no visual effect upon F/T cycling as described in ICBO AC24.

Table 3 - *Freeze/Thaw, AC-24 Section 6.5.1*

Sample Prep =	1. Cut 23 226 mm^2 square test panels in half in order to form a butt joint. 2. Apply WRC to surface of the test panels. 3. Seal the back and sides of the panels.
Testing =	1. Run 5 panels through 10 freeze/thaw cycles. 2. Each cycle: => 49° C for minimum of 8 hours. => total immersion in water at 21.1 °C. to 26.8 °C for 8 hours. => -28.9 °C for 16 hours.
Condition of Acceptance =	Failure is defined as surface changes such as cracking, checking, crazing, erosion, or other characteristic that may affect performance as a wall cladding. Failure is also defined as delamination, or indications of delamination between components. (Viewed with minimum of 5X magnification)

ICBO AC 24, Section 6.6.1: Water Resistance - Water resistance (Table 4) evaluates the performance of EIFS incorporating a WRC under severe high-humidity conditions. The Water Resistance test panels must be a minimum of 102 mm by 152 mm, and they are prepared in the same manner as those used in the freeze-thaw test.

Testing includes a minimum of three panels which are tested in accordance with the *ASTM Standard Practice for Testing Water Resistance of Coatings in 100% Relative Humidity* (D 2247), which is run at 38° C. Fourteen days exposure is required, and the WRC must show an "absence of deleterious effects from the 14 days of exposure to water".

The WRC based on the new acrylic binder passes the Water Resistance test, showing no visual effects upon exposure.

ICBO AC 24, Section 6.8.1: Water Vapor Transmission - Water Vapor Transmission (WVT) testing provides information about the WRC's ability to breath, that is, to allow water vapor to pass through the coating so as to not trap water inside the substrate. WVT is determined using the *ASTM Standard Test Methods for Water Vapor Transmission of Materials* (E 96), Water Method. WVT is reported in grams per square meter per 24 hours.

WRC films were cast on release paper using a 40 mil draw-down bar. The films were dried for 7 days at 25°C/50% RH. WVT was determined using a modified ASTM E 96, where a 6 cm. diameter metal can was used in place of the 'Perm' cup specified in ASTM E 96.

The WVT for the WRC/Adhesive is 37 to 44 grams per square meter per 24 hours (5 to 7 perms). Conditions of acceptance are that the WVT must satisfy one of the grade requirements in Table 14-1-A of UBC Standard 14-1 (see Table 5). This WRC/Adhesive meets Grade C, and D requirements.

Table 4 - *Water Resistance, AC-24 Section 6.6.1*

Sample Prep =	1. Cut 102 mm by 152 mm (min) test panels in half to form a butt joint. 2. Apply WRC to surface of the test panels. 3. Seal the back and sides of the panels.
Testing =	1. Run a min. of 3 panels for 14 days. 2. Test in accordance with *ASTM Standard Practice for Testing Water Resistance of Coatings in 100% Relative Humidity* (D 2247)
Condition of Acceptance =	The absence of deleterious effects from 14 days of exposure to water.

Table 5 - *WVT in gms/sq.meter/24 hrs.*

As listed in Table 14-1-A of UBC Standard 14-1:	Grade A	Grade B	Grade C	Grade D
Maximum	4	6	---	---
Minimum	---	---	---	35
Results of ASTM E 96 testing of WRC:		37 to 44		

ICBO AC 24, Section 6.10.1: Water Penetration Test – This AC-24 required test is listed for the record, it was not run in our laboratory because the size and complexity of the test exceed the physical capability of our laboratory. The Water Penetration test is conducted on a 4-foot by 8-foot (minimum) wall in which there are at least two vertical and one horizontal 1/8 - inch wide joints. The WRC is applied to the wall and cured. The assembly is then tested in accordance with *ASTM Test Method for Structural Performance of Exterior Windows, Curtain Walls, and Doors by Cyclic Static Air Differential* (E 1233), Procedure A, with 80% of the design load as the maximum load. The samples must be cycled through a minimum of ten cycles.

After testing in accordance to ASTM E 1233 the same assembly is then tested in accordance with *ASTM Test Method for Water Penetration of Exterior Windows, Curtain Walls, and Doors by Static Air Pressure Difference* (E 331). A 6.24 psf air-pressure differential must be maintained across the specimens for 75 minutes.

The condition for acceptance is that there must be no water penetration on the plane of the exterior-facing side of the substrate.

Our laboratory is not equipped to build and run this test assembly, therefore, we did not run the *Water-Penetration Test*. A commercial EIFS which includes a WRC should be tested for water-penetration as listed in Section 6.10.1 of the ICBO, AC 24 Acceptance Criteria. The EIFS manufacturer would provide details on application procedures including joint treatment and methods of flashing.

Summary of ICBO AC 24 Acceptance Criteria Required Test - Table 6 summarizes the performance of the WRC in the ICBO, AC24 required tests. The WRC based on the new acrylic binder passes the selected tests included in our study.

Table 6 *Performance of the WRC in ICBO AC24 Required Tests*

ICBO AC-24 Section	Test	AC-24 Requirement	Results for WRC
6.4.1	Tensile Bond:		
	plywood	Minimum of	pass
	mesh-faced gypsum board	> 15 psi	pass
	paper-faced gypsum board		pass
6.5.1	Freeze-thaw:		
	plywood	no surface	pass
	mesh-faced gypsum board	changes that	pass
	paper-faced gypsum board	effect performance	pass
6.6.1	Water-resistance:		
	plywood	absence of	pass
	mesh-faced gypsum board.	deleterious	pass
	paper-faced gypsum board	effects	pass
6.8.1	Water-vapor transmission of materials:		
	perms	UBC Standard	5 to 7
	gms/24 hrs/m^2	14-1, Table 14-1-A	37 to 44
6.10.1	Water penetration: (4 foot by 8 foot wall)	no water penetration	(not laboratory tested)

Testing to Further Characterize the WRC

We include several additional tests to further characterize a WRC intended for use in EIFS. *Percent Elongation* is included because it provides information about the ability of the WRC to maintain integrity during periods of movement, for example, at a joint in the substrate. *Resistance to Hydrostatic Pressure* helps assess the capability of the WRC to bridge cracks and maintain water resistive properties while doing so. *Accelerated Weathering* is used to determine if the EIFS composites loose strength when exposed to a combination of water spray and Ultra-Violet light. And finally, *Early Grab* provides a measure of the WRC's ability to resist EPS movement when a force is applied to the EPS after being adhesively attached using the WRC. Details regarding these tests, along with test results, for the WRC follow.

Percent Elongation - Table 7 shows the percent elongation at break for a WRC based on the new acrylic binder. Films were cast on release paper using a 40 mil draw-down bar and were then cured for one week at 25° C/50% RH. Films were tested at both room temperature and -17.8° C (0° F). *ASTM Standard Test Methods for Rubber in Tension* (D 412) was followed using a Tinius Olsen Model H10K-S Testing Machine with a crosshead speed of 1.25 cm (0.5 inches) per minute. Notice that the WRC provided over 75% elongation at room temperature, and that the elongation was not reduced by lowering the temperature to -17.8° C, a temperature typically encountered in many parts of the country.

Table 7 - *Percent Elongation at Break*

	Test Temperature	
	25° C	**-17.8° C**
Range:	75 to 130	80 to 115

Resistance to Hydrostatic Pressure - Table 8 shows the results of the Hydrostatic Pressure Resistance testing of the WRC/Adhesive. *ASTM Standard Test Method for Hydrostatic Pressure Resistance of Waterproofing Membranes* (D 5385) was followed. The test panel is a 23226 mm^2 (6-inch squares) pieces of substrate with 9 drilled 3.125 mm (1/8 inch) diameter holes. The test panel is cut in half, put back together to form a butt joint, and coated with trowel applied WRC. After a 48-hour room temperature cure, the seam in the test panel is expanded to 1.563 mm (1/16 inch), and the gap is maintained by carefully inserting spacers in the seam, making sure the WRC film integrity is not compromised by the placement. The test panel, with the expanded seam, is then placed in a hydrostatic test chamber (Figure 1) where water at a pressure of 15 pounds per square foot (psf) is applied to the surface of the WRC. This configuration is maintained for 48-hours. The panel is examined periodically to see if there is any water passing through the test panel/film. A "pass" means no water was observed on the back side of the panel during the 48 hour pressurized water exposure. The WRC based on the subject acrylic polymer showed no water penetration when tested over fiberglass mesh faced gypsum board or oriented strand board (OSB). Grade D building paper tested under the same conditions allowed water to pass through soon after the test was started.

Table 8 - *Resistance to Hydrostatic Pressure*

Substrate	Test Results
Mesh-faced gypsum board	**Pass**
Oriented Strand Board	**Pass**

Figure 1 - *Hydrostatic Test Chamber*

Accelerated Weathering - In order to check the accelerated weathering of this acrylic based WRC/Adhesive, we evaluated the tensile bond adhesion of EIFS composites incorporating the subject WRC before and after exposure in a Xenon Arc Weather-O-Meter.

Composite samples (triplicates) were prepared by gluing the substrate to a wooded test block, and gluing a piece of grooved EPS to a second wooden test block. After these assemblies cured, we trowel applied the liquid WRC to the substrate and used it to adhere the grooved EPS/wooden block assembly to the substrate. This composite was dried for one week, and then the edges and back were protected with an acrylic elastomeric coating. The composite samples were exposed in a Xenon Arc Weather-O-Meter for 0, 500, 1000, and 2000 hours. The groove in the EPS was placed such that water from the Weather-O-Meter spray cycle would drain down and through the channel. The configuration of the test sample placement in the Weather-O-Meter can be seen in Figure 2. The composite samples were tested under tensile using ASTM C 297, and the results are found in Table 9. Unexposed samples were included for comparison purposes.

The composite samples maintained tensile strength even after 2000 hour exposure in the Weather-O-Meter. When failure did occur, the mode was cohesive failure in either the EPS (Figure 3) or the substrate. These results attest to the durability of the WRC adhesive bond to EPS, plywood, mesh faced gypsum board, and paper faced gypsum board.

As of this writing, similar composite samples have been exposed outdoors in Spring House, PA, but have not been exposed long enough to report durability data.

Figure 2 – *Weather-O-Meter Exposure*

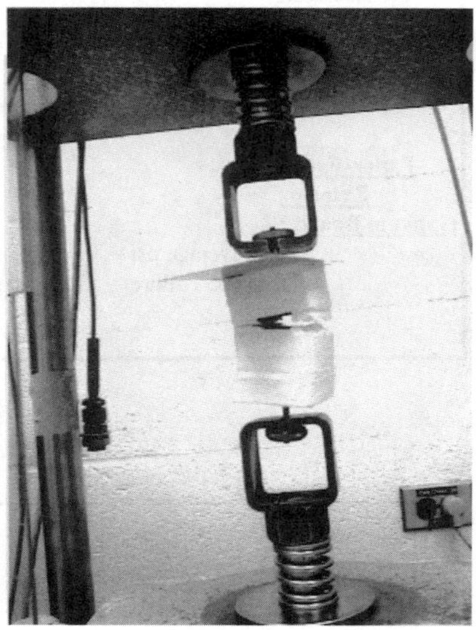

Figure 3 – *Tensile Testing*

Early Grab - A significant benefit of this WRC is that it can be formulated to provide early grab when it is used as the adhesive for the EPS board. This early grab is sometimes called "green strength" and provides wet grab which holds the EPS board when it is first set-in-place. Early-grab may allow rasping of the EPS board much sooner than the 24 hours needed when conventional liquid adhesives are used. As with conventional adhesives, the cure time depends on ambient weather conditions.

In the Early Grab test, a WRC/adhesive is trowel applied to a vertical plywood substrate. The EPS board is firmly placed onto the wet WRC/adhesive two minutes after the WRC/adhesive is applied. Laboratory weights are placed on the top surface of the EPS board at five-minute intervals starting 10 minutes after the EPS is put in place (Figure 4). The cure time needed to develop sufficient adhesive strength to hold 2000 grams (4.4 pounds) placed on the EPS board, without having the EPS board slide down the substrate, is recorded. Table 10 shows that a 1 foot square piece of EPS adhesively attached to plywood substrate with the new acrylic based WRC/adhesive can support 2000 grams after about 10 minutes A conventional latex EIFS adhesive required more than an hour cure.

Table 9 - *Tensile Bond Adhesion of Samples Exposed in the Weather-O-Meter*

Substrate	Hrs W-O-M =	0	500	1000	1500	2000
1/2" Plywood :						
	average psi =	33.7	31.8	31.2	34.9	35.3
	Type Failure =	EPS	EPS	EPS	EPS	EPS
Mesh-Faced Gypsum Board:						
	average psi =	29.3	35.3	35.3	19.8	27.8
	Type Failure =	EPS/ Substrate	EPS/ Substrate	EPS/ Substrate	EPS/ Substrate	Substrate/ EPS
Paper-Faced Exterior Gypsum Board :						
	average psi =	31.5	28.5	25.1	33.7	33.7
	Type Failure =	Substrate/ EPS	Substrate	Substrate	Substrate	EPS/ Substrate

Table 10 - *Early Grab Test Results*

Applied Adhesive	Time needed for 1' X 1' EPS square to hold 2000 gms
Acrylic WRC/adhesive	10 minutes
Conventional liquid adhesive	over 1 hour

Figure 4 – *Early Grab Test*

Conclusions

1. An acrylic emulsion polymer can be formulated into liquid Weather Resistant Coatings for use in EIFS cladding systems. Liquid weather resistive coatings are included in the ICBO, AC24 Acceptance Criteria For Exterior Insulation and Finish Systems.
2. Weather Resistive Coatings prepared with this acrylic binder can provide a unique rheology that allows them to also be used as the adhesive for the insulation board in the EIFS. The rheology manifests itself as a buttery feel under the trowel, and an "early grab" property that may shorten the adhesive cure-time needed prior to rasping the EPS.
3. Weather Resistive Coatings prepared with this acrylic binder meet the laboratory tested Acceptance Criteria as outlined in ICBO AC24 in Sections 6.4.1, 6.5.1, 6.6.1, and 6.8.1. The Water penetration test as outlined in Section 6.10.1 measures the performance of the full EIFS and was not performed in our laboratory due to the size and complexity of the test. This test must be performed on the complete EIFS in order for that system to fully comply with the AC 24 Acceptance Criteria.
4. Weather Resistive Coatings prepared with this acrylic binder provide exceptional elongation at both room and low temperatures.
5. Weather Resistive Coatings prepared with this acrylic binder demonstrate a degree of resistance to hydrostatic pressure, passing a 15 psf hydrostatic pressure test.

6. EIFS composites constructed with the WRC exhibited no loss in strength after 2000 hours of exposure in a Xenon Arc Weather-O-meter.

References

[1] ICBO, "Acceptance Criteria For Exterior Insulation and Finish Systems," AC24, dated October 1999.

[2] O'Brien, M. J., Burch, M. J., Rimmer, J. W., and McKelvey, J. L., "Durability of 100% Acrylic Polymers Used in Exterior Insulation Systems (EIFS)," *Development, Use, and Performance of Exterior Insulation Systems, (EIFS), ASTM STP1187*, M. F. Williams, R. G. Lampo, and R. G. Reitter, II, Eds., ASTM International, West Conshohocken, 1994.

[3] Lavelle, J. A., "Acrylic Latex Modified Portland Cement," *American Concrete Institute's Materials Journal*, Vol. 85, No. 1, 1988, pp. 41–48.

Author Index

Subject Index